holy **smoke**

incense & scent

in the home

holy **smoke**

incense & scent

in the home

Raje Airey and **Charlotte Melling**

photography by Craig Knowles

aqua marine

This edition is published by Aquamarine

Aquamarine is an imprint of Anness Publishing Ltd
Hermes House, 88–89 Blackfriars Road, London SE1 8HA
tel. 020 7401 2077; fax 020 7633 9499
www.aquamarinebooks.com; info@anness.com

This edition distributed in the UK by The Manning Partnership Ltd, 6 The Old Dairy, Melcombe Road, Bath BA2 3LR; tel. 01225 478 444; fax 01225 478 440; sales@manning-partnership.co.uk

This edition distributed in the USA and Canada by National Book Network, 4720 Boston Way, Lanham, MD 20706; tel. 301 459 3366; fax 301 459 1705; www.nbnbooks.com

This edition distributed in Australia by Pan Macmillan Australia, Level 18, St Martins Tower, 31 Market St, Sydney, NSW 2000; tel. 1300 135 113; fax 1300 135 103; customer.service@macmillan.com.au

This edition distributed in New Zealand by David Bateman Ltd, 30 Tarndale Grove, Off Bush Road, Albany, Auckland; tel. [09] 415 7664; fax [09] 415 8892

A CIP catalogue record for this book is available from the British Library.

PUBLISHER: Joanna Lorenz
MANAGING EDITOR: Judith Simons
EXECUTIVE EDITOR: Caroline Davison
DESIGNER: Louise Clements
STYLIST: Charlotte Melling
PHOTOGRAPHER: Craig Knowles
PRODUCTION CONTROLLER: Claire Rae

10 9 8 7 6 5 4 3 2 1

WARNING
Never leave incense or candles burning unattended. Neither the author nor the publishers can accept any responsibility for claims arising from the inappropriate use of candles, incense or other scented items featured in this book.

contents

introduction

Precious incense, as valuable as gold, was once the prerogative of gods and kings. The wealth of entire civilizations was built on the trade in aromatic herbs, resins and spices, while frankincense and myrrh, two of the most important incense substances, were gifts to the infant Christ, given in honour of his holy status.

To understand the value of fragrance we need to look back to the time when our ancestors discovered fire. Imagine the thrill of discovery as people realised that wood and plant materials could be dried and kindled to produce heat. Fire not only provided warmth, it also gave protection from wild beasts and an excuse to gather together for the telling of stories, the singing of songs, and the performance of dance, ritual and celebration as food from the successful day's hunt was prepared.

As different woods and plant pieces were burned, so different fragrances drifted up into the air, affecting mood and wellbeing; in fact, the word "perfume" comes from the Latin per fumen, meaning "through smoke". In those times, humans relied on their sense of smell for survival, recognizing which scents were safe and which spelled danger. From here it is a small step to see how a knowledge of fragrance began to develop, as people's finely attuned sense of smell distinguished between particular plants and their different effects. This knowledge was recorded by the shamans (medicine men and women) who understood which plants would heal, which would induce dreams or visions, and which would encourage relaxation and sleep. Because of their powerful properties, aromatic plants were honoured as gifts from the gods. Through the holy smoke, offerings and prayers were transported up to the heavens in fragrant spirals, consecrating the atmosphere and purifying the mind, body and soul of all those present.

Today, we live in an affluent age, yet many people have a sense of something awry. Clouds of poisonous smoke have billowed up to the heavens and we have lost touch with the natural rhythms of life. Looking for ways to create a better world and enrich our lives, we have turned to ancient cultures and back to nature for inspiration. The discovery of ancient traditions of using fragrance has been just one result of this search.

This book is a journey into the holy smoke, the world of natural fragrance. It shows how to use incense and aromatics as part of a modern lifestyle, knowing which scents to use to enhance different moods and for a myriad of different purposes, whether these be for healing, relaxation or love, to celebrate occasions, or to harmonize with nature's cycles. By learning how scent works and getting to know the subtleties of different aromas, it is possible to create a fragrant home that is both a delight to be in and reflects our own individuality. Home is where we live and work, and enjoy time with friends and family. It is our personal sanctuary where we relax, restore and refresh ourselves, physically, mentally and spiritually. Although most of us are familiar with using colour and décor to create a pleasing environment, the idea of integrating fragrance into the home may be new. Yet doing this is one way in which we can celebrate our own unique style and bring a fragrant dimension to our lives.

aromatics

Our society is visual and verbal. We place great store on appearance and to a large extent have forgotten the language of fragrance. But all this is changing as we rediscover the subtle and sophisticated art of aromatics.

Thousands of years ago, one way of extracting a flower's aromatic oils was by squeezing. In ancient Egypt, the blooms of exotic flowers, such as lilies, were gathered into a cloth bag. The bag was then twisted and tightly pressed until the fragrant oil oozed out of the petals.

fragrance in history

No one really knows the exact origins of where or when fragrance was first used. Yet since the earliest times incense and aromatics have been used in ritual and sacred ceremony, in medicine and healing, and to tantalize and delight the senses.

More than 100,000 years ago, the Neanderthals made offerings of pollen and flowers to their dead and early hunter-gatherer societies developed a knowledge of plants. Cave paintings found in Lascaux in France from around 18,000 BC show flowers and plants used for healing, while between 7,000–4,000 BC people combined olive and sesame oils with plants to produce fragrant ointments. Later, the use of fragrance is recorded in the earliest writings from the ancient civilizations of the Near and Middle East, China and India.

Around 5,000 years ago the Sumerians gave spiritual meanings to flowers and plants, and incense-burning rituals were linked to the planets. A little later, Babylonian merchants traded in aromatics, with around 200 varieties recorded on the city trade lists, including myrrh, cedar, frankincense and myrtle. Fragrance was used to communicate with the gods and tons of frankincense were burned each year on golden altars. In the Old Testament, God gave Moses detailed instructions on building an incense altar made from fragrant acacia wood and gold, together with recipes for perfumed anointing oil and incense using galbanum, frankincense, storax and onchya.

In ancient Egypt, aromatics were enjoyed for their medicinal and cosmetic purposes, as well as in sacred ritual. Fragrant resins, regarded as the tree's life-breath, were placed in graves to assist the soul's survival in the next world. Rites and prayers accompanied the creation of incense-burning mixtures. The recipes were recorded in "magic books" or on the walls of fragrance laboratories. Kyphi, meaning "welcome to the gods", was one of the most precious mixtures. It was used as an aid to sleep and dreaming and to treat all manner of complaints. Frankincense and myrrh were two key ingredients, while other substances such as cinnamon, spikenard, sandalwood, coriander, raisins and wine were added to make fragrant incense pellets.

above *In ancient China, camphor was a holy substance and the trees were planted at shrines and sacred sites. In Marco Polo's day, people paid for camphor in gold.*

below *The wealth of the legendary Queen of Sheba was based on the trade in frankincense and myrrh. Caravans of resin-laden camels were frequently hijacked by thieves.*

above left *Almost any absorbent material can be coated or infused with incense mixtures. In Tibet, rope incense is used in Buddhist temples. The smoke is believed to invite happiness and peace.*
above right *At one time, a plant's "essential" oils were believed to contain the elixir of life. The oils may be vaporized in an oil burner.*

Through trade and cultural exchange, the Greeks absorbed knowledge of fragrance from the Egyptians. They had just one word to describe the incense, perfume, spices and fragrant medicines they used so widely: *aromata*. Flat bowls made of iron, bronze and even gold were used as incense vessels and incense burning was used to assist prophecy and divination. Physicians such as Hippocrates and Dioscorides advocated the use of aromatics. The legendary megaleion, a salve containing cassia, cinnamon and myrrh, was used in the treatment of wounds.

The Romans, notorious for their lavish use of aromatics, used them to pamper and stimulate the senses. They layered fragrance onto their beds as well as their hair, bodies and clothes. The Emperor Nero had concealed pipes fitted into his dining rooms that sprayed fine mists of fragrance onto his guests, while panels slid aside to release showers of rose petals. Wealthy citizens literally bathed in perfume and enjoyed regular massage with scented oils. Delicate floral perfumes such as lily, violet and rose were especially popular.

Alongside developments in the use of aromatics in the Middle East and Mediterranean regions, the fragrant arts also progressed in India and the Far East. In Ayurvedic texts from India, incense was recommended in order to prevent infection, purify the environment and stimulate prana (the life force). Burning incense was believed to restore harmony to body and mind and increase goodness, or *sattva*. One of the most important remedies was sandalwood, a symbol of longevity. Hindu temples were often made from this exquisitely fragranced wood, and temples themselves were known as *gandhakuti*, houses of fragrance. The use of incense extended to love and pleasure. Written in the 4th century AD, the Kama Sutra, the classical Indian text on the art of love, included knowledge of fragrance as one of the sixty four arts that an educated man or woman should possess.

In ancient China fragrances were classified into six divisions, each associated with the mood or characteristic the aroma induced: luxury, tranquillity, beauty, nobility, refinement and solitude. The Chinese made extensive use of fragrance, not only using it to perfume themselves, their clothing, their homes and their temples, but even going so far as to print money on perfumed silk. They believed that perfume represented the elixir of life, and that fire had cleansing and purifying powers, with the smoke attracting good spirits.

In the **ancient world**, incense and fragrance permeated almost

The tradition of incense burning travelled to Japan with the advent of Buddhism in the 6th century. Buddhism had always used incense burning to support meditation and the journey of the soul to perfection. The Japanese used incense to refine and tame the senses and set up special schools to teach the sacred art of koh-doh or "the way of fragrance". Different incense was designed to support the soul at different stages of spiritual development and time itself could be measured in fragrance. Incense clocks changed their scent to mark the hour, while the seasons of the year were acknowledged with different perfumes. The Japanese made a distinction between *soradaki*, incense burning for pleasure and *sonae-koh*, incense offering to Buddha.

In Arabia, enthusiasm for fragrance permeated every area of life. Incense was burned to mark important occasions, such as birth, marriage or the signing of a legal contract, while the custom of offering guests a round of after-dinner perfumes is one that continues to this day. The heady scent of musk was reputed to be the favourite fragrance of the prophet Mohammed, and the substance was used in the building of mosques, its aroma to be released by the heat of the sun.

Above all else, however, it was the delicate scent of rose that came to permeate Moslem culture. Rosewater was used in a myriad of ways, from flavouring sweets to scenting prayer beads, while, according to legend, fountains of it could be found in the caliph's palace gardens. Poets, scientists, healers and scholars alike wrote about fragrance, and it was Avicenna, an Arab physician of the 11th century, who became the first to successfully distil rose essence. This rich, heady perfume, known as attar of roses, was brought to Europe by the Crusaders – and with it the secrets of perfumery.

As the Europeans eventually discovered, the Americas were also a rich source of fragrant treasures, used for centuries by its indigenous peoples in sacred ceremony and healing. In the Northern continent, Native Americans "smudged" the sick with bundles or braids of herbs such as white sage and sweet grass, while herbs such as tobacco and birch bark were put in a sacred pipe and smoked in ceremonies. In Central and South America, incense was widely used by the advanced civilizations of the Mayans, Incas and Aztecs. One of the most important incense-burning substances was copal. The Mayans considered it a gift from the gods, while the Incas dedicated it to their Sun god.

In Europe, Italy continued to dominate the fragrance trade, with Venetian scented gloves, stockings, shoes and even coins delighting the European elite. In the 16th century, Catherine de Medici carried the Italian influence to France, and towns like Grasse took the lead in the production of perfumed goods. Perfumed essences were sold to apothecaries who dispensed perfumes with their remedies. People believed that fragrance could protect from disease, which led to the use of pomanders, posies, strewing herbs and herbal fumigation.

During the 19th century the medical and cosmetic use of fragrance began to separate into distinct areas, as synthetically produced scents made them unsuitable for medicinal use. It was not until the 20th century that the French re-established the therapeutic use of fragrance with the discovery and pioneering of "aromathérapie", the modern term for this all-encompassing art.

right *Sandalwood incense has a warm, sweet, woody fragrance. Making clever use of the monsoon rains, Egyptian sea-faring merchants were able to import this valuable commodity from India. It was an important ingredient in Kyphi mixtures.*

every aspect of life, from **sacred ceremony** to everyday affairs.

attuning the senses

At one time, humans relied on their sense of smell for survival and were able to detect the most subtle nuances of scent. Today we are rediscovering the forgotten language of fragrance.

Our sense of smell is registered in one of the oldest parts of the brain, known as the limbic area. When we smell a scent, aromatic molecules are inhaled and waft through the nasal cavity to the olfactory (scent) membrane at the back of the nostrils. Here, millions of microscopic hairs gather and process "smell data". This causes chemical and electrical changes in the nerve cells, triggering messages that pass directly to the olfactory bulb in the brain to be registered as a particular smell. The olfactory bulb acts as a sorting house, organizing the scent signals into patterns or groups before they make their way into the limbic system, bypassing the rational, conscious mind. For convenience, perfumers divide these scent patterns into "aroma families". While there is no universal agreement about these, one fairly common division is into floral, citrus, herbaceous, woody, earthy, resinous and spicy. Some fragrances may fall into more than one family because of their complex chemical make-up.

While we know that our sense of smell is connected to taste, most of us are unaware that other senses may be involved. The Incas reputedly linked fragrance with colour, sound and even constellations in the sky, while the Japanese refer to smelling incense as "listening" to the fragrance. More recently in the West, the art of combining aromas has been likened to "writing music in smoke". In the 19th century, Piesse, a French perfumer, categorized aromatic plant oils according to a scale of top, middle and base notes, depending on how quickly they evaporated. The lightweight top notes are those that hit you when you first smell the fragrance. They stimulate and then disappear, leaving the heart of the "symphony", the middle notes that linger for several hours. Finally, the slowly evaporating base notes come to the fore, the deep chord of their scent lingering long after the top and middle notes have faded. Top notes tend to be green, citrus, or light florals, while the middle notes are mostly floral, with maybe some spicy and woody tones. The voluptuous base notes are earthy and resinous, but also with woody and spicy tones. According to this theory, a harmonious blend is created when all three levels are balanced, having a roundness of "volume" and a coherent theme. Fragrance is subtle and complex, built up of a layering of scents.

opposite *In theory, we are able to recognize tens of thousands of different scents, yet our sense of smell is relatively unrefined. Learning to appreciate the different qualities of fragrance families is one way of reattuning our sense of smell.*

above and below *Incense sticks, cones and spirals usually contain a blend of different incense powders and essential oils. As your sense of smell begins to reawaken, it is possible to detect the "layering" of different fragrances in the incense.*

FRAGRANCE FAMILIES

floral	*sweet, flowery scents, ranging from delicate to heavy and sensuous*	lavender, rose, geranium, neroli, violet, jasmine, tuberose, honeysuckle, ylang ylang
citrus	*energizing and refreshing scents*	bergamot, mandarin, orange, grapefruit, lemon, lime, citronella, petitgrain, lemongrass
herbaceous	*fresh "green" herbal scents*	camomile, rosemary, marjoram, lavender, clary sage, sweet grass, white sage, juniper tips, thyme
woody	*scents of the forest*	sandalwood, agarwood, cypress, cedar, pine, juniper berry, tea tree, eucalyptus
earthy	*musky and mysterious scents*	patchouli, rose mallow seed, spikenard
resinous	*distinctive, generally strong aromas*	frankincense, myrrh, copal, galbanum, benzoin, dragon's blood, sandarac, storax, dammar, labdanum
spicy	*scents ranging from warm and mellow to piquant and refreshing*	clove, cinnamon, cardamom, cassia, star anise, galangal, ginger, coriander, nutmeg

what is incense?

Incense is the fragrant aroma that is produced when certain plant material is burned or vaporized. It is also the term for the actual material – typically resins and barks, herbs and spices, and oils.

The word incense comes from the Latin *incensus*, meaning to kindle or alight. Aromatic substances are left to smoulder over a heat source, such as a charcoal block, or rolled and moulded into incense sticks, cones or spirals and set alight. The smoke drifts up into the air, permeating the atmosphere with its fragrant aroma and influencing mood and wellbeing.

Incense can be made from a variety of materials. In the past animal scents such as ambergris, civet and musk were highly sought after, but today plant-based substitutes are favoured in place of animal ingredients. In fact, the vast majority of incense material has always come from the plant kingdom. Almost any fragrant plant constituent can be used to make incense, including woods, barks, roots, leaves, fruit and flowers, as well as resins, gums and oils. These may be dried and used raw or made into pastes and powders. Semi-precious stones may also be added to the mix, while a sprinkling of gold dust makes for a truly special mixture.

There are literally hundreds of potential plant ingredients for making incense, but the tree resins, such as frankincense, myrrh, copal and benzoin, are among the most important and are often used as a base in incense-making. Also known as gum-resins, balsams and balms, these sticky substances either occur naturally or are produced by chipping away at the bark until the tree secretes a milky substance. This gradually solidifies into a white, golden or even black resin on exposure to the air.

Two of the most valuable incense woods are sandalwood and agarwood (or aloeswood). Sandalwood is a useful base in many incense formulas, as well as a burning agent in its own right. The best-quality sandalwood comes from the Mysore region of India. It takes more than 25 years before the wood's fragrance develops, with the heartwood yielding the sweetest fragrance, although this is very rare and extremely expensive. Agarwood, known as *jinkoh*, or "the sinking wood" in Japan, is probably the most expensive ingredient in perfumery. The scent has been compared to a blend of sandalwood and ambergris. The fragrance is formed after the wood has been infected by fungus, causing the tree to produce resin. The more resin the wood contains, the more aromatic the fragrance, and the heavier (and more prone to sinking) the wood becomes. Agarwood is usually burned as tiny slivers the size of a grain of rice.

above *Incense sticks are not always small. It is possible to find gigantic burners such as these, which work really well outdoors.*

left *There are endless permutations when it comes to making incense. The ingredients may be crushed into powders, rolled into sticks, or shaped into cones or spirals. They may also be used "raw" as loose resins, dried leaves or herb bundles, or left as tiny splinters of fragrant bark.*

1 *Assemble all the ingredients, keeping them separate from one another and working with each ingredient one by one.*

2 *Grind the ingredients into powders with a pestle and mortar. Start with the dry ingredients and leave the resins until last.*

Whatever plant part is used, however, the fragrance of incense is derived from the aromatic oils that occur in all parts of the plant tissue. These oils represent the life-blood of the plant because of their "essential" nature, each having a unique fragrance and character. It is these essential oils, or "concentrated smells", that form the basis of modern-day aromatherapy. Some of these oils, particularly exquisite florals such as jasmine and rose, are extremely costly, owing to the sheer volume of plant material required to produce the oil. For instance, it takes approximately eight million jasmine blossoms to make just 1kg (2lb) of "absolute". The intoxicatingly fragrant, white flowers are traditionally hand-picked at night because that is when their perfume is at its strongest. Originating in India and the Arabian Gulf, jasmine is now cultivated in Morocco, Egypt, India, Italy and France.

Making your own incense

The advantage of making incense yourself is that it allows you to experiment with fragrances that you particularly enjoy and design your own custom-made blends. Create fragrances with a clean, invigorating aroma to greet the day, seductively sensual scents for a romantic evening, calming mixtures for relaxation, or exotic eastern blends for entertaining friends. Working with the raw ingredients also increases your sensitivity to their unique aromas. To begin with, try to include at least one resin or wood in your mixture as a base, with maybe one or two herbs.

loose incense

The starting point in all incense-making is to turn the raw ingredients (the resins, barks, herbs and so on) into powder form, which is suitable for mixing and burning. Although it is possible to buy many ingredients as powders, sometimes you will need to grind them up yourself. To do this you need a heavyweight pestle and mortar and/or a spice or coffee grinder. Any equipment you use should be dedicated to incense-making only.

It is best to pulverize the dry ingredients first, leaving the resins until last as they are the most messy to work with. Putting the resins in the freezer for 15–20 minutes first should make them much easier to handle. Woods can be difficult to powder, and an old-fashioned hand-crank coffee grinder may be the best choice. Once the ingredients are powdered, mix them all together, adding the resins to the dry mixture. If you are using any essential oils or powdered gemstones, sprinkle them in last. If you are not using the incense straight away, store it in a dark, glass, screw-topped jar. Label it, giving the date you made it, and a list of the ingredients. This "loose" incense may be burned over charcoal or used as a basis for making other types of incense.

incense pellets

For richness and depth of fragrance, you can take your incense making a step further by adding binding ingredients to the loose incense to make "incense pellets", known as *nerikoh*, or incense

3 *Mix all the powders together, adding the crushed resins. If you are using essential oils, sprinkle them in at the end.*

4 *To burn the incense use a suitable fireproof container and some incense charcoal. Light the charcoal and sprinkle on the incense.*

balls in Japan. Binding ingredients include honey, raisins or other dried fruit, and pliable resins, such as labdanum. Soak the dried fruit overnight in a little red wine to soften. Add enough binding to the loose incense to make a sticky paste and knead well. Form the dough into small balls or pea-sized pellets and spread them onto a wooden board or wax paper to dry. The drying time can take up to four weeks depending on the climate. The fragrant pellets are burned over charcoal, or, as an alternative to burning, they may be placed in decorative sachets and used to scent clothes, rooms and furniture.

incense cones and sticks

When we think of incense, most of us think of sticks or cones because these are the varieties that are most readily available in the shops. Most manufactured incense contains a chemical burning agent, usually saltpetre (potassium nitrate). Incense can also be made using a natural burning agent called makko (also known as tabu), which comes from an Asian evergreen tree. Some of the finest quality, handmade incense uses makko and it is recommended that you use this when making your own incense.

One of the secrets of making incense cones or sticks is to make sure that your ingredients are ground into an ultra-fine powder. Let the ground mixture sit overnight in order to allow all the smells to blend together. The makko is then added to the loose mix. You will have to experiment with how much makko to add because the amount varies depending on the composition of the mix, with mixes that are high in resins requiring more makko. As a rough guide, for mixtures that contain no resins, use 15ml (1 tbsp) of makko for every 75ml (5 tbsp) of loose mix, increasing the makko content up to 60ml (4 tbsp) for mixes high in resin. Slowly add a little water and mix well, using your hands to knead the mixture. It should become gummy and pliable, yet still hold its shape. On a piece of wax paper, form the mix into free-form cones or roll into sticks and leave to dry, away from a direct heat source. If the incense burns slowly or erratically, then you will need to add more makko; if it burns too quickly, then decrease the makko content. Do not worry if the cones or sticks do not work out the first time. You can easily break up the incense and start all over again, adding more makko or loose incense mix until it burns perfectly.

INCENSE STARTER KIT

resins	woods	herbs and spices
benzoin	agarwood	cinnamon
copal	(also known as	cloves
dammar	aloeswood or *jinkoh*)	lemongrass
frankincense	cedarwood	jasmine
mastic	cypress	lavender
myrrh	pine	rose
sandarac	sandalwood	white sage

methods of burning

Incense comes in all shapes, sizes and colours, but there are two basic types. Combustible incense can be burned directly, while non-combustible or loose incense must be burned on top of a suitable heat source.

Combustible incense is the sort that is most readily available in the shops. You can buy it in the form of cones, sticks, spirals and other shapes. It usually contains potassium nitrate (saltpetre) to help it burn, although it is possible to find combustible incense without this chemical ingredient. Once lit, it will burn at a steady rate. With the exception of joss sticks, which are made from pure incense paste, incense sticks are usually made from bamboo skewers coated with powdered incense. Like joss sticks, incense cones are made from pure incense paste, except the paste is pressed into cone-shaped moulds rather than through thin holes. Incense spirals are made in a similar way to joss sticks, except the incense paste is moulded into spirals rather than sticks. To use combustible incense, simply light one end, fan out the flame to make sure it is alight and allow it to burn slowly of its own accord.

Non-combustible or "loose" incense is made from raw ingredients that have been ground into a fine powder and mixed. As it will not burn by itself, loose incense is usually smouldered on an incense charcoal block. Loose incense tends to produce more smoke and needs more attention while burning. Incense charcoal blocks are usually round or rectangular and generally contain saltpetre, a chemical burning agent, to help them burn. They should be kept wrapped in aluminium foil and stored in an airtight container. Alternatively, powdered makko or bamboo charcoal, which do not contain chemical burning agents, may be used. To reduce smoking, loose incense may be burned on a mica plate placed over the lighted charcoal or makko in a fireproof container. To light the charcoal, hold it with tweezers and suspend it over a candle flame until it begins to glow. Sometimes the lit charcoal is buried in sand or ash in the incense vessel, and the mica put on top. The secret with burning loose incense is to use a little at a time.

holders and containers

For loose-incense burning, you will need a fireproof container (or censer) in a bowl or saucer shape. If you do not want to use charcoal, incense stoves (similar to essential oil burners) use a lighted candle to produce the heat. For combustible incense, choose a suitable holder with a groove or bowl to catch the falling ash.

clockwise from top left *Incense burning bowls are perfect for loose incense. They are available in a variety of designs. Some are open, while others have holes in the lid for the smoke. Special cone-shaped burners are also available for incense burning. They also have holes to let the smoke through. One of the easiest ways of burning incense is the stick method. A block of wood or a piece of stone also makes an unusual holder and ash catcher for an incense stick. Special holders are available with a tiny hole for the stick and some kind of recess to catch the ash.*

Bring traditional pot pourri up to date with modern arrangements and containers. Dried cones, leaves and plant materials are offset by the colour and texture of white pebbles.

other scenting methods

Aside from incense burning, there are countless ways of using natural fragrance to create mood and ambience around the home. These include fresh and dried flowers and plants, and many ways of using essential oils.

One of the most popular and versatile ways of creating fragrance is with essential oils. These are volatile, concentrated substances extracted from many different types and parts of plants and trees. They are highly aromatic and are used in perfumery as well as in modern-day aromatherapy. They may also be used in incense preparations. Essential oils are quick and easy to work with as they do not need any complicated preparation or much in the way of equipment. They also make an excellent choice for those sensitive to smoke. Always purchase pure, unadulterated oils and take care when handling neat oils as they can cause skin irritation.

For an instant and subtle effect, try vaporizing essential oils. The scent molecules are rapidly dispersed through the air so that you are hardly aware of breathing them in. The fragrance may be used to purify the atmosphere, repel insects, clear the air of cooking smells or simply for enjoyment, depending on which oils you choose. There are many ways to vaporize essential oils. Electrical vaporizers are the safest type to use, and a good choice in busy areas such as the hallway. Lamp rings are also safe and becoming more widely available. However, the standard ceramic "oil burner" is still by far the most popular method. The burner is designed to hold a night-light candle positioned beneath a small reservoir of water to which a few drops of the essential oil/s are added. As the water evaporates, the room becomes suffused with fragrance. When choosing a burner, select one with a generously sized reservoir so that the water does not evaporate too quickly. Never leave lit burners unattended. To remove the sticky residue of burnt oil from the water tray, use an alcohol-based substance such as surgical spirit or neat vodka.

For a quick burst of fragrance, aromatic mist sprays have a variety of uses. They are useful for freshening up stale air or to give a subtle fragrance around the whole house. Alternatively, you may like to mist around yourself to freshen up your aura (energy field), or to use the fragrant mist when steam-ironing, avoiding synthetic fabrics or ones with special finishes. Fill a spray bottle with water, add a few drops of oil and shake well. As a guide, use up to 15 drops of oil to 150ml (¼ pt) water. Spritz as needed, avoiding polished surfaces. Once you've used essential oils in a spray bottle, do not reuse it for any other purpose. Always shake the bottle before use to disperse the oil droplets.

above *When choosing essential oils, make sure the fragrance is 100% pure and has not been synthetically blended. Essential oils are available in a wide selection of scents from all the fragrance families.*

below *Candlelit essential oil burners remain one of the most convenient ways of scenting the home. The oils disperse into the air, leaving no trace of smoke or ash.*

right *Heat and light are essential for life on earth. Whether or not we realise it, candlelight stirs deep in our ancestral memories, a reminder of our connection to "higher powers" or the light of life.*

fresh and dried materials

If pot pourri conjures up memories of your grandmother, think again. Bowls of fading rose petals left to gather dust are giving way to contemporary chic as we adapt this traditional technique for scenting the home. Collections of natural objects, such as feathers, pine cones, pebbles and stones, or pieces of bark or driftwood, can be scented with essential oils to make fragrant and decorative arrangements. Even if you decide to opt for a more traditional pot pourri then look for unusual or stylish containers to keep it in. Lidded bowls help to preserve the fragrance and keep out the dust – just take off the lid when you use it. It's also a good idea to try making your own mixtures. Many commercial preparations are artificially coloured and use harsh, synthetic perfumes to scent the dried wood and plant material. Use a mixture of dried petals, leaves, herbs, spices and barks, adding orris root to help preserve the scent and, finally, pure essential oils. Remember that the secret with any scented arrangement is to keep it fresh and clean.

Dried fruits and herbs can also be used to enjoy scent around the home. Citrus pomanders made from dried, spiced fruit may be enjoyed for their fragrance or used to deter moths and insects. Oranges, lemons, limes or kumquats are studded with cloves and rolled in a mixture of ground cinnamon and orris root. The fruit is then left to dry in a warm place for up to six weeks, depending on the size of the fruit. The pomanders can be hung in cupboards or used as festive decorations. Orange pomanders work particularly well in winter, although well-made pomanders will keep their scent for years. Alternatively, dried bunches of herbs such as lavender or rosemary can be hung in cupboards, used in herb pillows and sachets, or as a decorative detail – fixed to photo frames or chair backs, hung above a mirror, or as a welcome on the front door.

If you are lucky enough to have an open fire, then you can experiment with using aromatic woods and dried plant bundles on the fire. Fragrant woods include apple, laurel, pear and cherry, while some of the best dried plants are angelica, lavender, rosemary, pine cones, juniper twigs, lovage and sage.

If you do not have an open fire, then soft flickering candlelight can set the scene for a relaxing evening. Scented candles are available in the shops, but make sure these use genuine essential oils and not harsh synthetic scents. You can also scent your own candles. Light the candle and let the wax melt a little around the wick, then blow it out. Add two or three drops of essential oil to the molten wax pool and relight the candle. Keep the wick trimmed short otherwise the flame will be too big and the aroma will not last long. Essential oils are highly flammable, so always extinguish the flame before adding the oil to the molten wax.

Finally, no home is ever complete, either inside or outside, without fresh plants and flowers – the original source of natural fragrance. It takes only a few garden roses in full bloom to impart a delicate scent to a room or an arrangement of lilac blossoms to make an unforgettable statement. Other fragrant flowers from the summer garden include herbs, mock orange blossom, wisteria, alyssum, lavender and white lilies. Spring bulbs such as lily-of-the-valley, hyacinth and narcissi are also strong fragrance favourites. If you do not have a garden, some scented plants, such as alyssum and wallflowers may be grown in containers. Others, such as white jasmine, gardenia and heliotrope, may be grown as indoor plants. Rather than making random choices about which of these to choose, consider carefully the effect you want to achieve, bearing in mind the colour and shape of the plant and how it fits in with your décor as well as its unique fragrance.

right *Herbs from the kitchen garden make unusual indoor displays. When creating plant arrangements, consider how the different scents will blend together to create a pleasing aroma.*

Essential oils are extracted from a plant's leaves, flowers, fruits,

stems, roots or bark to create an explosion of fragrance.

the ambient home

Walk into someone else's home and it is almost guaranteed that one of the first things you will notice is how it smells. Make incense and fragrance your greatest ally to create a welcoming ambience in your home.

living rooms

The living room is usually one of the largest and most important rooms in the home. Traditionally the showpiece of a house, it is the place to make a statement about yourself and your lifestyle. Complement the interior style of each room with carefully chosen fragrance.

above and opposite *Choose incense burners and fragrance to harmonize with your lifestyle. Woody fragrances such as sandalwood or rosewood work well in this light, open room with its emphasis on rich browns and dark woods.*

Living rooms are multi-functional spaces. They are used by all members of the family at different times of the day. It is also where guests are entertained – ranging from an informal "drop-in" visit by close friends to organized and more formal occasions. However, the living room is also the place where you spend most of your leisure time. It is where you sit and flop at the end of a tiring day, maybe to watch TV, listen to music, read a book or the newspaper, or maybe just to sit and chat about the day's events. For this reason, the room needs to work in terms of comfort and practicality as well as style and design flair.

When using fragrance in the living room, you first need to decide which particular activity will be going on, as this will influence your choice of scent. During the day, children may appreciate fun and fruity fragrances such as watermelon, strawberry, apple, peach or coconut. But for general daytime use, a safe bet is to opt for light floral or citrus fragrances. Geranium has a light, uplifting aroma – the Victorians were so fond of it that they placed tubs of the fresh plant at strategic points around the home so that when women brushed by their long skirts would rub against the plants and release the scent. The scent of the incense or essential oil, however, is quite different to the fresh flower; it is more subtle and delicate, with a slightly minty aroma. Violet also has a delicate floral smell, and combines well with geranium or rose. Citrus scents are particularly welcoming on a hot summer's day or as a pick-me-up in the late afternoon when energy levels are flagging. Lemon has a clean, fresh aroma, while the distinctive scent of bergamot is useful for eliminating stale, musty odours and cooking smells.

left *Incense is more than a fragrance. It is also a colourful accessory that can be used to co-ordinate with your colour scheme. Here, soft pastel pinks add softness and a feminine touch.*

It is the leaves from the bergamot plant that give Earl Grey tea its flavour and scent, yet you need not restrict yourself to drinking scented tea. Loose tea also makes a good choice for incense burning. The floral aroma of jasmine tea or the smoky scent of lapsang souchong will add an extra dimension to your tea drinking and socializing. Or you could try scented fruit teas such as peach, mango or passion fruit. Even unscented green or black teas work well. If you do not want to burn the tea, then you could arrange it in tiny open bowls for it to release its subtle fragrance.

If you are having a stressful day, take fifteen minutes out of your schedule to light some incense. Burning incense encourages you to slow down and catch your breath. It reawakens and refines the senses and can enable you to move on again with renewed energy. Make sure the doors and windows are closed and turn off any music, the radio or television. Sit quietly in your favourite armchair or sofa and let yourself relax as the fragrant smoke drifts through the room. Sandalwood, jasmine or any of your favourite fragrances should do the trick.

As the evening draws in you can afford to be a little more adventurous with your fragrance choices, opting for deeper, warmer notes or heavier, resinous scents. The soft, sweet scent of vanilla is ideal for a quiet, relaxing evening, and it is also comforting after an upsetting day. Its warm, mellow tones particularly suit large, light, spacious areas. When entertaining, burning incense can set the scene and make your guests feel relaxed and welcome. For something rich and exotic choose musky, earthy scents such as patchouli or the deep resinous scent of frankincense. Or for a lighter, more laid-back approach choose blends containing sandalwood or delicate florals such as rose, jasmine, lavender or orange blossom. During the long nights of winter, spicy aromas such as cinnamon, cloves or cardamom work well.

Add a touch of decadence to clean whites with shots of purple and pink. A perfect space to relax and enjoy.

right *The incense and candles add another dimension to this beautiful living room. Delicate floral scents such as rose, neroli or lavender would all work well here. Or for something a little more exotic, try the voluptuous and heady scent of vanilla or ylang ylang.*

In the past, the servants of the wealthy left "perfume cakes"

Traditionally, the fireplace was the focal point of the living room. Before the invention of central heating, an open fire was the heart and soul of the home, providing warmth and a comfortable place where friends and family could gather together to talk, relax, read or tell stories. Although most modern homes no longer have an open fire, many living rooms still have some kind of fireplace area which tends to be used as a focal point for arrangements of flowers and plants and other precious items. You might like to try fresh flowers or elegant bowls of sweet-scented bulbs such as narcissi and Easter lilies. The fireplace is also one of the more obvious places in which to burn incense, either in stick holders or in special incense burners. If you do not have any sort of fire area, then you can create a symbolic one using candles, their flickering flames acting as a gentle reminder of the warmth and light of an open fire.

If you are lucky enough to have an open fireplace, then it's definitely worth taking the trouble to use it, especially on cold winter nights. The sound of dry timber cracking and hissing and the mesmerizing sight of the vermilion flames licking up the chimney are relaxing and a delight to watch. Add another dimension to the experience by using aromatic plant material. You can look for fragrant woods by collecting the branches of fruit trees such as apple or cherry, or fallen pine cones and pieces of bark. At the end of the summer, collect herbs such as lavender, sage, rosemary or thyme from the garden, and dry them indoors, away from direct heat. You may want to use some of the herbs in your incense-making or they can be tied together and made into herb bundles for throwing on the fire. Their fresh scent will soon clear away any lingering cobwebs in the atmosphere.

Alternatively, you can toss incense cones or sticks directly into the fire. Woody scents such as cedar, pine, cypress or any other evergreen tree are especially grounding and their fresh invigorating aroma reminiscent of the forest. Or, to follow a nomadic custom which is still practised in the Arabian Desert, you could throw pieces of frankincense resin onto the fire as a welcome to your guests. If you want to use essential oils, add a few drops of oil to pieces of dry timber and leave for a while before putting on the fire. In fact, it is best to let the woods absorb the oils for about 15 minutes before setting them alight. In winter a combination of ginger, sandalwood and orange should help get you and your guests in the celebratory mood. Never add essential oils directly to the fire as they are highly flammable.

left *An open fireplace is an excellent place in which to burn some incense, either in stick holders or special incense burners. In this rustic, earthy living room, a large piece of timber has been used to create the perfect incense holder in an open fireplace. The rough-hewn chair with its co-ordinating cushion continues the woody, natural theme. When not in use, an open fireplace is also ideal for positioning arrangements of fresh, scented flowers such as lilies or bowls of fragrant pot pourri.*

to smoulder gently on the fire to **freshen** the air in a room.

dining rooms

From an informal family supper to a lavish dinner party, the skilful use of fragrance can enhance the enjoyment of delicious food and good company. Incense can be used to harmonize with the mood and theme of the occasion.

Eating is an opportunity for members of the household to get together. It is also a good excuse to meet up with friends. Whether informal or formal, remember that good eating is not about spending hours slaving away in the kitchen. The important thing is to relax and have fun, enjoying the delicacies of the food and conversation in a leisurely way.

The idea of a separate room for eating is becoming a thing of the past. To save on space, dining areas are often part of another room such as the kitchen or living room. They may also have to double up as a work or study area. This can produce a conflicting array of aromas and a backdrop of clutter which should be cleared away before eating. The best dining areas are ones that are light and spacious with accessories kept to a minimum and where fragrances blend in a co-ordinated way. The main focus of attention is the table. Table settings may be formal or informal, depending on the type of meal and occasion. A dining environment can be enhanced by subtle fragrance, but it must complement rather than compete with the food.

Tiny incense cones not only smell good but also look very attractive when used as table decorations. They are probably best left unlit until after the meal is over, when you can light one or two at a time. Cones are available in all colours and scents. Light fragrances such as apple blossom or almond are soft and inviting, while stronger, spicier scents such as cinnamon, ginger or cassia may be enjoyed at more formal dinner parties.

Candles, of course, are a traditional dining accompaniment and there are many varieties available in the shops. Low-level, nightlight-style candles look pretty, particularly when arranged in multiples of three or four. For subtle fragrance, add a drop of essential oil to the molten wax in a few candles, remembering to blow out the flame before adding the oil. Geranium is delicate and refreshing, while juniper or lemongrass will help to clear the air.

above *When planning a dinner party menu, the delicate aromas of the different foods should harmonize. A light vegetable soup not only smells appetizing but looks very tempting when served in dark-glazed bowls.*

below and opposite *Coloured ginger incense cones make attractive table decorations. Group them in holders that compliment your tableware. Light them once you have finished eating.*

left These foil-wrapped nuggets look good enough to eat, but they are, in fact, blocks of amber resin. Unwrap them and use as room or drawer scenters. You could also give them out to your guests at the end of a dinner party.

Pure beeswax candles are fabulous. They are more expensive than ordinary candles but well worth it for a special dinner party. The creamy yellow wax has a very faint honey-like scent. This can be enhanced with a woody essence such as sandalwood or, if you are at the pudding stage, a more exotic floral such as ylang ylang might be more tempting. There are many different types of candleholder available, although some of the nicest for table decoration are made out of glass and metal.

The dining table has always played host to fresh flowers. Carnations, summer pinks, roses, mimosa or freesias are traditional choices that give a sweet and refreshing scent. When using flowers for the table, think in terms of less being more. One or two stems in a beautiful vase can have far more impact than a complicated arrangement. On the other hand, for Roman-style decadence and a heart-warming welcome scatter fresh rose petals across the table.

Throughout the year, fruity scents lend a pleasing note. Bowls of fresh fruit create a sense of abundance, while dried-fruit pomanders make for an interesting table centrepiece. Clove-studded oranges work well at a winter dinner party, particularly when combined with evergreen leaves. Alternatively, the dried zest of oranges, lemons, limes and grapefruit may be dropped into bowls to form a quick pot-pourri arrangement. Combine with sandalwood powder or chippings and add a few drops of essential oil to strengthen the fragrance.

Chic, cool understatement creates the perfect backdrop for eating and entertaining in style.

At the end of a meal it is customary to pass around sweets such as mints, chocolate, ginger or pieces of Turkish delight. Mint and ginger are digestive aids, bitter chocolate a taste sensation, and rose a gift from the heart. The sweets usually accompany a hot drink. Scented teas such as jasmine or mint or tiny cups of black coffee are traditional fare. You could even go a stage further and follow a traditional Arab custom and give your guests a round of after-dinner perfume. Mist sprayers containing light floral waters made from orange, lavender or rose make a refreshing end to a sumptuous banquet.

right The clean white lines of this dining area are suited to both formal and informal dinners. Keeping the windows free from window hangings allows the beauty of the garden to almost become part of the dining room. On the table, settings are kept pure and simple. Dried vanilla pods in a ceramic container make an intriguing centrepiece. Golden bowls of foil-wrapped amber resin introduce an air of decadent luxury and opulence in an otherwise minimal setting.

kitchens

The hub of domestic life, the kitchen is one of the busiest areas of the home. It is where food is prepared and cooked and is a natural den of aromas. Make sure these are welcoming and seductive rather than unpleasant and off-putting.

Kitchens need to be warm and comfortable, practical yet inspiring. Of course, the kitchen is more than just the workroom of the house. It is also where friends drop in for a chat and where guests inevitably seem to gather at parties, perhaps to be near the food and drink, but also because there is generally something warm and inviting about this room, the heart of the house. Keep it clean and smelling fresh and it will be a pleasure, rather than a chore, to spend time there.

Judging food by its smell is one way to tell whether it is fresh and appetizing. For aromatic culinary heaven, herbs and spices must surely win the day. Keeping them out on display reminds you to use them and gives ample opportunity to sample the different aromas. Prettily shaped star anise has a warm, aniseed-like fragrance that can restore flagging energy levels, while greyish-green cardamom pods, widely used in India and the Middle East, are said to stimulate sexual desire. Even staple culinary ingredients such as black pepper, ginger and garlic each have their own unique aroma-properties: both black pepper and garlic are fortifying, while ginger is energizing and grounding.

When you can get them, fresh herbs are generally far superior to dried in cooking, imparting a fuller flavour as well as a more tantalizing aroma. It is difficult to resist the piquant smell of fresh coriander or the pungent uplifting scent of basil. The kitchen windowsill is the perfect spot for displaying herbs, but take care not to over water. Many common garden herbs such as lemon mint, parsley, chives or sage, as well as seasonal salad leaves are fairly easy to look after. The bright green leaves look stunning against the light and release a delicious whiff of fragrance onto your fingers when pinched.

All pictures *The kitchen is home to a mouth-watering array of aromas. Freshly ground coffee and dark bitter chocolate are a classic combination. Here, the fragrance is enhanced by the sweet smell of vanilla pods and incense sticks.*

above left *Candles can be used to good effect in the kitchen to create a soft, relaxing atmosphere. Arrangements of candles grouped together in blocks look far more effective than a single flame.*
above right *Do not keep fresh fruits hidden away, but put them out on display. Citrus fruits, such as limes and lemons, add a bold splash of colour and give off a fresh, zesty scent when cut open.*

The mouth-watering smells of oven-baked bread and sizzling bacon mingle together over a leisurely Sunday breakfast. Or the heavenly aroma of freshly ground coffee and dark bitter chocolate is guaranteed to tantalize most people's taste buds. The kitchen is home to a myriad range of enticing aromas from foods and drinks. The secret is to contain those smells and not have them permeating into other unwanted areas for hours afterwards. Adding cloves to a simmering pan of hot water will act as a natural air freshener, the clear and spicy vapours clearing the atmosphere.

Perhaps more than any other room, it is especially important to keep the kitchen clean and free from stale odours, which can indicate the presence of harmful bacteria. Many fragrances have cleansing and antibacterial properties, especially pine and citrus scents such as lemon, both of which feature in many commercial cleaning products. A good blend for the kitchen is a combination of eucalyptus, cinnamon and clove or pine, grapefruit, lemon and thyme. You could also use any of the fragrances singly. For a chemical-free cleaning product, add a few drops of the pure essential oil to the rinsing water or onto a sponge. Or to give cutlery drawers and crockery cupboards a subtle scent, tuck aromatic cotton balls in the corners.

Whether chopping vegetables, washing up, ironing, cleaning shoes or repotting house plants, you want to make sure the kitchen is a pleasure to be in. Lighting some incense as you work can inspire you so that positive thoughts go into what you are doing. It is even said that food prepared while in a positive state of mind has greater nutritional value than meals put together under stress and negativity. If you burn incense while preparing food, take care not to do it too close to the working area; burn it in a far corner of the room. Alternatively, place bowls of loose incense pot pourri in appropriate spots. Warm, spicy fragrances such as cinnamon, nutmeg, vanilla and clove coupled with orange will add a rich mellow tone and create nourishing vibrations to keep the atmosphere free from unpleasant odours.

right *Many kitchens have some kind of dining area. Clean lines and a light, open space create a relaxed and welcoming atmosphere. Here, splashes of bright pink suggest a sense of fun and easy living in contrast to the purity of the white. Fragrance the air with scented candles in geranium, orange spice or cinnamon.*

Bold colours bring life and vitality to a backdrop of neutrals and

whites. Sharp citrus scents keep the atmosphere clean and fresh.

bedrooms

More time is spent in the bedroom than in any other single room. The bedroom is a private sanctuary, a peaceful place of rest and renewal, and an intimate space for lovemaking and sensuality.

The bedroom is one of the most important rooms of the home. It should be a tranquil sanctuary where restorative sleep and intimate exchange can take place in a relaxed way. It is said that when it is laid out and styled appropriately it brings health and harmony to the whole family. For a bedroom to work, it is essential that anything connected with the outside world is kept to a minimum and out of sight. This means putting shoes and clothes away and never having anything to do with work in the bedroom. Apart from essential items, electrical appliances are best kept somewhere else. To create a relaxing ambience, keep furniture and clutter to a minimum and choose décor and furnishings carefully to reflect your style.

Centre-stage is the bed area, so take care not to let other pieces of furniture compete for attention. For a contemporary look, low-level beds with a metal or wooden frame work well. A table at the foot of the bed creates an extra dimension to the bed space, providing a display space for personal effects. Scatter cushions and bed linen can be used to great effect, co-ordinating the colour to harmonize with the walls and window hangings. Clear aqua colours and deep blues are easy to live with, while warmer pinks and reds are perfect for love and intimacy. When choosing your bedroom colours, go for ones that you will not tire of too quickly.

Incense can be colour co-ordinated to match the room. One or two sticks out on permanent display look attractive and can, of course, be burned at any time. There are many ready-made incense fragrances on the market with highly evocative names. If you prefer pure rather than blended fragrance, however, then scents such as lemongrass or lavender are fresh and uplifting. Sandalwood is always a good choice for unwinding; many commercial blends contain sandalwood as a base.

above *Scented candles compliment the refreshing turquoise and white colour scheme in this bedroom.*
opposite *For a fresh striking look, splashes of clear turquoise add a shot of colour to white, bringing lightness and a sense of freedom – wonderful for waking up to each morning.*
below *Incense sachets filled with fragrant powders or dried herbs have always been popular for scenting linen and clothes. Co-ordinate the fabrics with the rest of the bedroom.*

right *Scented candles that include dried flowers and herbs not only smell good, but look interesting. Lavender is a popular fragrance choice for the bedroom as it has a relaxing and cleansing effect.*

For a less visible approach, scenting clothes and linen has been popular since antiquity. The Japanese and Chinese tucked sachets of perfumed powders into the sleeves of their kimonos, while rose, violet and patchouli were used to scent the imported Indian shawls made popular by Napoleon's wife, Josephine. Tiny scented pillows are available in the shops or you can make your own using dried herbs or incense pellets. Use them in cupboards or drawers for a subtle fragrance. For scented lingerie, vanilla, rose and jasmine are alluring choices, while lavender is a classic choice for linen. Certain aromas will also keep moths at bay. These include cedar, clove, lemongrass, patchouli and lavender.

The best bedrooms are cocoons of comfort. A rich sensuous haven will set the scene for intimacy at the end of a long day. Deep warm colours and soft textures can be used to great effect, complemented by low-level lighting. Deep reds, pinks or mauves, with maybe a splash of orange, are perfect colours for snuggling up at the end of the day. Take care not to overdo strong bright colours as these are energizing and may disturb restful sleep.

Legend has it that Mark Antony had to wade through a carpet of rose petals to get to Cleopatra's bed. For a more subtle approach, use fragrant powders, herbs or oils on bedroom floors. To freshen up carpets and rugs, sprinkle them with baking soda, adding essential oils to the dry powder. If you like it, patchouli has a deep, sensuous, earthy scent, while palmarosa has a fresh and slightly sweet, rosy aroma. Alternatively, sprinkle powdered incense in a fine mist over the carpet or place dried herbs under a rug so that the fragrance is released when it is stepped on. When tired of the scent, you can clean it away.

Strongly scented plants and flowers are best avoided in the bedroom as the fragrance can be too overpowering, especially last thing at night. On the other hand, scented candles, some incense or bowls of fragrant powder can help create a romantic mood. Florals such as rose, jasmine or neroli are all good scents for enhancing lovemaking. Other incense resins, herbs or oils can be used to encourage sleep. These include lavender, camomile, hops, valerian, marjoram and galbanum.

left *There are many ways of storing incense. Ceramic containers such as this one are very contemporary. The lid doubles up as an incense holder and ash-catcher.*
right *Delicate touches of lilac and soft maroon bring understated luxury to the light woods and cream in this tranquil bedroom. Dried bunches of lavender flowers give off a subtle fragrance.*

bathrooms

Scent heaven. From a quick, energizing shower to a long, relaxing soak in the tub, the bathroom is an oasis of fragrance, home to an enticing array of scented products used in pampering bath-time rituals.

The bathroom has risen from its humble origins to become one of the most luxurious and desirable rooms of the home. In fact, many a home has been sold on the basis of its bathroom. Because the room is generally small, it is possible to experiment with different materials and design schemes that would prove prohibitively expensive elsewhere. Warmth, plenty of natural light and good ventilation provide a good platform to work from.

A white bathtub or shower is classic and will stand the test of time. Keep designs simple and avoid overly fussy or complicated patterns. Setting the bath in some kind of recessed surround creates a display space for incense or scented candles while open shelves provide useful storage for towels and scented bath-time products. Positioning scent bags between the clean laundry will give them a delicate fragrance.

The bathroom can give away many telltale clues about you. Many bathrooms have a lingering scent reminiscent of their owners, for it is here that perfumes in one form or another are in liberal use, applied directly to the body or to the bathwaters. But this has not always been the case. In the Victorian era, scent was never used directly on the body, as its effects were considered too dangerous. Some scents, such as musk, were even banned.

Fortunately, contemporary bath-time culture owes more to the Romans who spent hours each day at the public baths, socializing as they enjoyed soaks and steams followed by massage with scented oils. Bathing is much more than a functional activity. It is a ritual act designed to cleanse and refresh the body while restoring inner harmony and balance. Aromatics play a key role in this ritual, different fragrances being used in different products for a variety of purposes, but all to be enjoyed for their "feel-good factor" as their aromas waft through the steamy atmosphere.

opposite and above *The contrast of white with dark browns make this bathroom fit for a king. Coloured incense sticks in white containers add a splash of colour and scent.*

below *The natural theme is echoed with these coconut-shell candle holders. Coconut has a sweet, creamy fragrance and combines well with tropical fruits such as pineapple and banana. Many bathtime products contain these smells.*

There are countless ways of using fragrance in the bathroom. Burning incense is one way. Vaporizing essential oils in a burner is another. A few drops of essential oil may also be added directly to a warm bath or onto a shower sponge, although take care if you have sensitive skin, as essential oils are highly concentrated and can aggravate some skin conditions. If that is the case, floating scented candles in the bath may be a better option.

For top-to-toe body treatments, there are hundreds of different types of scented products available, designed to cater for every occasion and budget. A vast repertoire of scented soaps, shampoos, shower gels, bath crystals and milks are used in the first stages of the bathing ritual. The next stage is pampering and re-moisturizing the skin with body lotions or oils, using a nourishing cream for the delicate facial area. A light dusting of powder and a delicate mist of perfume or toilet water are the finishing touches. When choosing scented products, look for high-quality ingredients and let your nose be the guide. You can also make your own products using unscented cream or oil bases, adding essential oil blends to the mix. Remember that scents work in layers or notes, so choose your fragrances carefully so that they work as a symphony and do not jar.

Since Roman times, the clean and refreshing scent of lavender has been one of the most popular bathroom fragrances. In fact, the word "lavender" is taken from the Latin *lavare*, meaning "to wash". It is ideal for a relaxing evening bath, particularly when combined with camomile; the combination is exquisite. To make a change from essential oils, use the dried herbs, either in bundles under running bath-taps or in herb sachets (or even herbal tea bags) in the water. Scents such as eucalyptus, bergamot or pine are deodorizing and head clearing, while ylang-ylang or gardenia are exotic and sensuous. Gardenia and jasmine can also be grown indoors as houseplants. The bathroom, of course, is a good place to keep plants, particularly those that enjoy moisture such as ferns.

above *Glycerine soaps are available in a wide variety of colours and scents. As they are semi-opaque they look wonderful when displayed in front of a natural light source. Here, the green of the avocado and the orange of the mandarin complement one another perfectly. Avocado is nourishing for dry skin, while orange is a good cleanser for oily skins.*

right *Scented granules make a colourful and fun alternative to more traditional pot pourri. Here, they are artistically arranged in an enticing palette of colours and smells. Names such as summer rose, juicy lime and ocean breeze conjure up the mood of each different fragrance.*

hallways

More than just a passageway from one room to another, the hallway is the gateway to the home, its main entrance and exit area. Make first and last impressions count using subtle fragrance to enhance and beautify the space.

It does not matter whether the hallway is large or small. The important thing is to create an open, inviting space that draws you in to the rest of the home. This can be achieved through light and colour and by keeping the space clutter free. A display table for a lamp and fresh flowers or incense is simple and effective. Pictures, photographs and personal mementos are often displayed in the hallway, but make sure they work in terms of the overall design of the space and do not turn it into a random collection of disparate objects.

If you have a dark or small hallway, use bright or light colours and cheerful fragrances such as orange, bergamot, mandarin or geranium to lighten up the space and lift your spirits. Burn them as incense or vaporize essential oils. To create the welcoming smell of home, warm and spicy aromas such as cinnamon, cassia, star anise and clove are good winter choices, while a profusion of intoxicating summer blooms, such as stocks, lilac, sweet peas or cabbage rose, can permeate the whole house with their fragrance. Freesias too have a strong scent and are pretty to look at. Fruity scents such as strawberry, pineapple or melon are fun, and an ideal welcome for party guests.

Hallways also need to be robust enough to withstand the busy traffic of people coming in and out. Choose colours and floor coverings that work for your lifestyle. There is no point in having a pale carpet if it is going to get heavy wear from children and pets. Similarly, fragrance choices need to work with their surroundings. A clean minimalist look will benefit from a purity of scent. Agarwood is reputedly one of the finest fragrances in the world and is used in Japanese incense; sandalwood too has an uplifting, inspiring aroma. For something more down to earth, a blend of vetiver, myrrh, sandalwood and patchouli will create a calm, dignified entrance area, the rich scent lingering once the last guests have departed.

above *Scented candles float in a little water amidst a scattering of rose petals. Delicate florals lend a subtle nuance of fragrance to this light, airy hallway.*
opposite *In parts of Asia and the Far East no entrance is complete without some burning incense. The wisps of smoke purify the atmosphere and embrace all who enter and leave the building.*

home offices

More and more people are working from home. Depending on space, the home office may vary from a separate room to a designated workspace within another. What matters is that the space is a pleasure to work in.

The best workstation is one where everything is within easy reach. Basic office equipment includes a desk and chair, computer, printer, telephone and fax. Depending on the type of work, extra worktop space may be needed for a scanner or photocopier and somewhere for drawing or design boards. Plenty of natural light, clean and simple lines and a tidy workspace help to keep the mind focused; it is difficult to concentrate amid a sea of paperwork. To stay in control, prioritizing tasks and being ruthless with throwing things away is essential.

It is worth spending fifteen or twenty minutes at the start of the working day to focus, making a realistic plan of what you can achieve. As you do so, burn some incense. It will clear the atmosphere from the day before and give a fresh start for the day ahead. A good morning blend is a combination of dammar, mastic, elemi and lemongrass. The transparent lemony fragrance is clarifying and supports inner balance.

Potted plants and quartz crystals are useful in the office as they can help absorb the energy waves emitted by electronic equipment. It is a good idea to take a break away from the computer at regular intervals to recharge your energy levels. Certain fragrances can also help minimize the effects of electromagnetic stress. A combination of peppermint, geranium and rosemary is said to be effective.

There are many fragrances that have specific head-clearing properties and are useful when engaged in mental work. If you like the smell, the sharp, penetrating aroma of scents such as eucalyptus, peppermint or camphor clears muddled thinking and is an antidote to daydreaming. A little less intense but as effective are herbal aromas such as rosemary or basil. In ancient Greece, students wore garlands of rosemary to aid concentration, while basil stimulates memory and sharpens the intellect. Other scents for mental fatigue and loss of concentration are coriander, lemongrass and pine. Or when you keep finding excuses to nip away from the desk, try grapefruit.

top *A pot of basil works well in the office as its pungent aroma helps to sharpen the mind. The loose pine in the chic, black bowl also improves the powers of concentration.*
above *Peppermint incense cones help clear the mind for creative thought.*
opposite *A rosemary plant is a good choice. Its refreshing fragrance helps counter mental burn out and protects you from computer fatigue.*

outdoors

Outdoor areas add an extra dimension to the home. Whether you have a large, country garden or a small, urban courtyard, they provide a special place for relaxing and entertaining and a sanctuary for flowers and plants.

When planning an outdoor space, look for visual harmony with shape, texture and colour and use fragrance to add an extra dimension. The most obvious place to start is with fragrant shrubs and plants. Some of the most popular varieties include sweet alyssum, wallflowers, night-scented stock, nicotiana (tobacco plant) and, it goes without saying, many varieties of rose. The trumpet-like white flowers of the datura plant have an intoxicating, heady scent, although take care as the plant is poisonous. White lilies have a similarly heady fragrance.

At first, you may not think of using incense outdoors. Yet the garden is also a good place to experiment, particularly if you are sensitive to smoke. Some of the best incenses for using outside are the extra-large sticks which can be pushed into the soil in a pot or flower bed to stand up between the plants. You can experiment with different aromas depending on the season. For instance, florals such as jasmine, rose or lavender are perfect in mid-summer, while in late summer/early autumn, the spicier, earthy scents of cinnamon, nutmeg or patchouli, or woody scents such as rosewood are more fitting.

To create a space for sitting, some shelter and shade is essential. Some kind of terrace next to the house is usually the best area for entertaining and enjoying the rest of the garden. When well designed, it becomes a natural extension of the building, rather than something separate. Choose colours and furniture that blend with indoors. In the evening, subtle lighting creates a warm, inviting glow. Candles, flares or outdoor lanterns will create a romantic atmosphere on the terrace. To keep annoying insects at bay, there are several scents that are particularly useful. These include tea-tree, eucalyptus, lavender, rosemary, thyme, lemon grass, citronella and cinnamon.

If you do not have much of a garden, cheat and use window boxes or tubs. Aromatic kitchen herbs such as rosemary, thyme, sage, parsley, hyssop, mint or chives look fantastic in terracotta pots. The white of a disused porcelain sink or enamel bathtub is a good contrast for brightly coloured geraniums and other summer bedding. For window boxes, sweet alyssum has clusters of small white blooms and a strong sensuous scent, while the more delicately scented candytuft suits an urban environment.

top *The beauty of nature is recreated in this waterlily candle. Like the lotus flower, waterlilies are a traditional symbol in the East of spiritual unfoldment and wisdom.*
opposite *Having easy access to an outdoor area adds another dimension to the home. Large incense burners help to keep insects away.*
below *A scented candle will create a soft and fragrant aroma on a warm summer's evening.*

fragrant moods

Freshly mown grass, bitter chocolate, a favourite perfume – even the mention of a particular smell has the power to conjure up memories and feelings. Scent and its powers of association can be used to enhance wellbeing in many different ways.

refresh

Wake up your senses. Light, uplifting fragrances restore flagging spirits and are an antidote to feeling sluggish. There are many scents with refreshing properties. Combine them with bursts of clear, fresh colour to set the mood.

Some of the best colours for lifting the spirits and lightening a dull mood are vibrant greens, acid yellows, sharp turquoises, and icy aquas. Clear white, of course, always looks clean and cool. This does not mean that you have to go overboard and base a whole room scheme around these colours, but rather use them as vibrant splashes where appropriate. Cushions, rugs, window hangings or a vase in the right place can all be used to refresh the senses.

In terms of scent most of the citrus fragrances are a good first choice. The zingy aromas of lemon or lime are excellent pick-me-ups, especially first thing in the morning. A few slices of the fresh fruit in hot water or drunk in herbal tea make a good substitute for the usual early morning cup of tea or coffee. Lemon has a clean, lively scent which stimulates the body into action, while lime, although not quite so sharp, is nevertheless stimulating and refreshing. The fresh smell of grapefruit is also awakening. The raw fruit is popular at breakfast time, giving a delicious and healthy kick-start to the day.

Clean citrus fragrances are particularly helpful when working or studying. Grapefruit is good when your confidence has taken a knock. It helps overcome obstacles and provides a boost to get back on track. Lemon sharpens the mind and clears confusion, while lime is good for nervous exhaustion and stress. The slightly sweeter, warmer scent of lemongrass is useful for mental fatigue and loss of concentration, as is the invigorating fragrance of petitgrain. Petitgrain comes from the leaves and twigs of the bitter orange tree. Its sharp orange smell is effective for sharpening the mind and preparing the intellect for further work.

above *Lemons and limes are tangy and refreshing. Their freshly squeezed juices are wonderful in hot weather.*
opposite *The icy aqua colours of the two vases are stimulating. A green, pine-scented incense coil evokes the coolness of a forest glade.*
below *To restore flagging energy, a quick burst of any citrus fragrance in an oil burner usually helps.*

above left *The vibrant lemon and green streaks in this huge white lily are echoed in the incense and burner alongside. The fragrance being burned is lemongrass.*

above right *A slice of lemon or lime added to hot water makes an instant fruit tea that is a refreshing pick-me-up.*

Of all the citrus fragrances bergamot is one of the most beguiling. Along with petitgrain, it is one of the main ingredients in the classic eau-de-Cologne toilet water. Its scent is derived from the peel of the ripe fruit and is what gives Earl Grey tea its distinctive aroma and taste. It is an effective deodorizer and one of the best scents for eliminating stale, musty odours and cooking smells. Similarly, the deodorizing properties of both lemongrass and citronella also make them effective room fresheners.

Refreshing scents are not confined to the citrus fragrance family. The fresh smell of pine needles is cooling and enlivening, while the smoky, woody aroma of cypress trees is cooling and calming. Like the citrus smells, both pine and cypress are also deodorizing and a counter to mental fatigue. However, take care when using these fragrances as they can be overpowering. They both blend well with citruses such as lemon and bergamot.

Other refreshing tree scents include eucalyptus and tea tree, both native to Australia, and Borneo camphor from Indonesia. All three have a penetrating aroma that smells slightly "medicinal", making them useful for refreshing the atmosphere in a sick room and clearing away mental cobwebs. Borneo camphor, referred to as *hon-sho* in Japan, is milder and more pleasant than other more common forms of camphor, a great deal of which is synthetically produced. It is the only form of camphor that is safe to use as an essential oil, which is sometimes referred to as borneol.

Of course, many refreshing aromas find a natural home in the bathroom where they are widely used in soaps, scrubs, shampoos and cleansers, as well as in toilet waters and aftershaves. Lemon has cleansing, astringent properties and is widely used in products for greasy skin and hair. Grapefruit is a good detoxifier and works well in body scrubs, adding zing to a tired complexion. In toiletries for men, lighter citrus scents are generally preferred with subtle fragrances from the forest adding a deeper, mellower note.

From the kitchen garden, there is a wide variety of culinary herbs with a refreshing, uplifting scent. These include parsley, lemon balm, rosemary, fennel and peppermint. The delightful citrus fragrance of lemon balm is irresistible to bees and lemon balm honey has been valued since ancient times. The menthol aroma of peppermint is cooling and refreshing, while peppermint tea is a staple drink in many hot countries. The slightly hot, spicy fragrance of parsley is particularly useful for counteracting the smell of garlic and onions, while rosemary has a mouth-watering aroma. Fennel has a sweet aniseed-like aroma. Sprinkled with lemon or lime juice, a drizzle of olive oil and baked in the oven, fresh fennel makes an appetising accompaniment to a main meal.

Choose **refreshing** and uplifting **citrus** scents such as lime,

above *This refreshing pot pourri is scented with lemongrass, petitgrain and citronella. Pot pourri is useful as a natural room freshener to keep the air smelling clean and fresh. These scents work well in the bathroom or kitchen.*

lemon, bergamot or grapefruit for added zest and zing.

relax

In the busy, modern world, fragrance really comes into its own. There are many different aromas that have a relaxing effect on the brain and nervous system, encouraging emotional balance and a quiet mind.

It is difficult to relax when all your attention is focused on the world "out there" and the mountain of things that has to be done. Yet for health and harmony it is important to be able to switch off and to "just be". Being unable to relax is linked with stress and insomnia as well as many other common health and lifestyle problems. Neutral colours and soft lilacs, blues or mauves can help to create a calm, relaxing environment, bringing the focus away from the hustle and bustle of daily life. You need to find the right balance when using shades of purple and blue, however, as too much can become heavy and depressing.

A warm, leisurely bath before bedtime is one way of letting the cares of the day wash away. There are many fragrances that can be used to support a relaxing evening bath in preparation for a good night's sleep. Use them as incense or vaporize as essential oils, a few drops added directly to the bath water or to an oil burner. Floating scented candles are another option. Like its colour, lavender is soft and delicate and a powerful relaxant. It has been a favourite scented herb for thousands of years. The Romans were fond of using it to scent their baths, the fragrant water cleansing and relaxing both body and soul. Lavender eases away everyday stresses and strains and helps tight, tense muscles relax. A tea brewed from fresh lavender heads was reputedly a favourite drink of Queen Elizabeth I of England who drank it in copious amounts to relieve her frequent migraine headaches. Spanish sage is also good for relieving tension headaches, the symptoms of stress and exhaustion. It has a fresh piney-lavender scent.

above *Place a herb bag under your pillow as an aid to relaxation and restful sleep. Lavender, marjoram, hops or camomile are all suitable fillers. Use the dried flower-heads.*

below *The soothing scent of lavender is a good antidote to stress. Infuse a few dried sprigs in distilled water to make a lavender-scented splash.*

The deep, earthy scent of vetiver is grounding and relaxing. It is useful for calming the nerves and easing tension and anxiety, particularly when feeling unable to cope under pressure. It is a natural sedative and can encourage a good night's sleep.

left *Lavender's blue-violet flower-heads look pretty when dried and arranged in open bowls. Widely regarded as one of nature's greatest "cure-alls", lavender has a balancing effect on moods and is one of the most popular fragrances.*

Camomile is another gentle, delicate floral that encourages relaxation and restful sleep. The aroma is said to resemble fallen apples and has been valued since antiquity for its sedative powers. It was traditionally used in sleep pillows for a child's bed, while at one time camomile lawns were very fashionable – the more the flowers were walked on the stronger their fragrance. Camomile is widely drunk as a calming, relaxing tea, while bunches or sachets of the fresh or dried herb can be floated in the bath or used in incense blends. Neroli is another exquisite floral with sedative and relaxing qualities. Moving away from the florals, men may prefer the smoky, earthy fragrance of vetiver, widely used in the manufacture of men's toiletries. Related to lemongrass, vetiver is deeply relaxing and helpful for insomnia, stress and tension.

Other relaxing, sedating herbs include hops, valerian and marjoram. All three have an antispasmodic action, relaxing tension in tight muscles and calming an agitated nervous system, although it has to be said that the smell of valerian is generally unpopular. It's probably best combined with other suitable herbs, such as passionflower, hops and lime flowers, and drunk as a tea. Like camomile, hops are often used in sleep pillows to treat insomnia, although in large amounts they can have a stupefying effect. Along with marjoram, hops are said to quell sexual desire in men, although in women they reputedly have the opposite effect. Dried marjoram has a warm spicy fragrance when burned and combines well with myrrh and mastic in incense blends for relaxation.

Burning incense can help you switch off and relax at any time of the day. Sit back and enjoy the aroma.

There are many other fragrances for relaxation, although frankincense, myrrh and sandalwood are some of the most effective. Rapid, shallow breathing is one of the signs of stress, but frankincense slows and deepens the breath. As an experiment, sit with closed eyes for ten minutes, light some frankincense and imagine tension being released each time you breathe out. Myrrh has a musty aroma and is helpful for agitated states and for cooling heated emotions. It is grounding and calming and ideal for when you need to come back down to earth. For tensions caused by insecurity sandalwood is a good choice. Its sweet woody aroma combines well with many other fragrances to produce a calming blend that helps steady the nerves.

right *The delicate fragrance of this incense stick is a blend of sandalwood, rose and passionflower. In India and Tibet, sandalwood is used to treat tension headaches and to ease emotional weariness. Both rose and passionflower have a calming effect on the nerves. Rose calms overactive personalities who cannot switch off, while passionflower is a natural sedative and helpful for emotional burn-out and work-related stress.*

energize

Sometimes getting away from it all is what is needed to invigorate the spirits. But this may not always be possible and solutions need to be found closer to home. There are many fragrances that can help restore vitality.

Feeling tired and run down is part of life, especially after the winter, a long period of hard work or after illness. Fiery reds, vibrant oranges and hot pinks are good colours to use when energy levels are low or when you want to brighten up a dull day. Rich purples, dark blues and too much black, grey or brown are best avoided, as these can drain your energy even more. Similarly, scent should be used with care, steering away from heavy fragrances that linger in the atmosphere for hours. Some of the lighter, fresher and spicier aromas, on the other hand, are stimulating and fortifying, and can help put some of the zest back into life.

Many everyday herbs and spices are warming and energizing and can either be included in your diet or used in incense blends or essential oil preparations. Fiery ginger is used extensively in Chinese medicine to warm and energize the body. Chinese herbalists regard it as being high in yang energy, the outward-going and active principle needed to get things done. Its warm, fresh and spicy aroma is a stimulating tonic. Closely related to ginger is galangal, widely used in Tibetan and Ayurvedic medicine. Galangal strengthens and stimulates the flow of energy through the body. Its gingery, camphorous fragrance works well in incense blends. In medicinal preparations, ginger is sometimes combined with camphor to improve the circulation and warm up the body.

Black pepper, one of the earliest known spices, is also warming and strengthening. It has a sharp, hot scent that blends well with other spices, such as cinnamon or cloves. Cinnamon is warming and offsets the sharpness of black pepper, while the hot, spicy scent of cloves simultaneously lifts and adds depth to the fragrance. Alternatively, athletes use muscle rubs containing black pepper and rosemary to help them achieve faster running times and to reduce muscular fatigue.

opposite and above *The spicy scent of ginger combines with orange in these incense cones. Ginger not only warms the body, but sharpens the mind and restores emotional vitality. Orange counters gloomy thoughts.*

below *The distinctive aroma of camphor is invigorating. Camphor has many medicinal properties. Its energizing effect stimulates the body's healing powers.*

right *The reds and oranges in this pot pourri mix give off a spicy sweet aroma that is both relaxing and energizing. The woody scent of copaiba balsam is used as a base for vanilla, cinnamon, cloves and a touch of black pepper.*

When listless, coriander will boost energy and recover a lost appetite. Coriander has a warm, provocative scent that can encourage a tired mind into action, and instil the feeling that life can be handled a step at a time. The plant's seeds are used in incense preparations and to produce coriander essential oil. The ancient Chinese believed the seeds contained the power of immortality. It is widely used as an incense-burning substance in many Arab countries where its stimulating fragrance is believed to lift the spirits and restore inner balance.

Feeling vital depends on clear thinking, strong motivation and a sense of purpose as well as physical energy. When these disappear it's easy to succumb to a "can't be bothered" feeling. Mental fatigue and apathy call for scents that will blow away the cobwebs and restore a positive outlook. Fresh, uplifting herbaceous fragrances can clear confusion. Again, many of these are in regular culinary use, but they can also be enjoyed in incense and essential oil preparations.

Rosemary has a camphorous odour that is strengthening, refreshing and centring. It is an excellent restorative and good for exhaustion, muscular aches and pains, and mental fatigue. Traditionally, rosemary was said to improve memory. Perhaps this is because the body is encouraged to take in more oxygen, thereby increasing the flow of blood to the brain. Like rosemary, basil also stimulates sluggishness, improves concentration and sharpens the memory. The aroma is refreshing, but can be overpowering so use in moderation. Peppermint also has a refreshing scent that is energizing and head-clearing. Both the ancient Romans and Greeks wore garlands of it to encourage mental clarity. Peppermint can restore lost confidence and is good for when energy levels are low. Together with rosemary, it can stimulate inspiration. If you are feeling really stuck, a blend of rosemary, peppermint and a little black pepper should help to motivate you.

There are also certain resins which have an energizing effect. Copaiba balsam has a woody, spicy, slightly peppery fragrance, similar to patchouli. It is stimulating and energizing and dispels exhaustion. It blends well with vanilla, whose warm, sweet aroma is a mild stimulant, offering encouragement to make an extra effort. The elemi tree grows in the tropics of Asia. The resin has a spicy, tenacious aroma with a fresh lemony, herbaceous overtone. Its clarifying, energizing properties stimulate mental work and alleviate mild depression. Elemi combines well with other fresh lemony scents such as lemongrass and mastic. Mastic resin, also known as pistachio resin, is light to pale yellow in colour. It has an uplifting effect and will help counteract a lethargic feeling, particularly in the winter months.

left *For an instant energy buzz, look out for strong reds and pinks. Combine them with hot, spicy scents for some warmth and cheer. This red, cinnamon-scented candle creates a warm glow that will lift flagging spirits.*

above *This pot pourri contains sandalwood powder, cinnamon and cloves. These combine to produce a delicate, warming fragrance that will energize body, mind and soul. It is the perfect combination when feeling drained and low.*

cheering

Everyone has down days now and again. Finding ways to cheer yourself up and not take life too seriously is one of the secrets of happiness. Have fun with lively and uplifting scents to create a joyful atmosphere and banish the blues.

Happiness is one of the most elusive of moods. No one has all the answers as to what makes someone happy, but negative thinking, unrealistic expectations and an inability to be in the present are all associated with unhappiness. One of the reasons that scent is so powerful is that it brings us into the here-and-now. There are also certain aromas that specifically counter negative thinking and promote an optimistic outlook and inner contentment. All the citrus fragrances are uplifting, while the light, delicate florals are a reminder of summer days. They can be complemented by the use of cheerful colours such as sunny yellows, vivid oranges, bright pinks or iridescent greens. Make the most of natural light and avoid dark or heavy colour schemes as these can become depressing, particularly in the winter.

Oranges and lemons seem to have an almost child-like innocence. They are associated in one way or another with some of the most cheering scents. Sweet orange has a fresh, uplifting aroma that helps disperse negative thoughts and revives the spirits. This succulent juicy fruit enlivens the winter fruit bowl and is a traditional ingredient in mulled wine. The Chinese traditionally offered gifts of oranges at New Year as symbols of happiness and prosperity. Mandarins (tangerines) have a more delicate flavour and refined aroma than oranges.

Bitter (or Seville) oranges, traditionally used to make marmalade, are also associated with two fantastic fragrances. Neroli, sometimes referred to as orange blossom, takes its name from a 16th-century Italian princess who is said to have discovered this haunting fragrance. Its sweet, floral aroma helps reduce fear of the future and dispel negative thoughts. Cousin to neroli, petitgrain is derived from the trees' leaves and twigs and consequently has a woodier, more robust scent. Petitgrain is particularly good for feelings of disappointment and loneliness and for overcoming the "winter blues".

below *The word for fish also means "surplus" in Chinese, and a goldfish is traditionally a symbol of luck and prosperity. Its bright orange colour is also cheering and optimistic. Alongside the fish bowl, these miniature scented candles introduce an element of fun. They have a fresh, fruity fragrance, resembling a cross between watermelon, apple, orange and tangerine.*

These scented shapes are an unusual form of pot pourri. They have a fresh, zingy fragrance which, combined with their bright colours, serves as an instant pick-me-up. Children will love them, but make sure that they do not think they are sweets.

right *Hundreds of years ago, nutmeg was widely used as a tonic for happiness. Its warm and provocative scent has a fresh, spicy top note. Nutmeg should be used in moderation. It blends well with mandarin.*

Lemon is a cleansing tonic, its lively scent bringing a reminder of warm sunshine. The crisp fragrance of lemon is also strongly present in the herb, lemon balm. In medieval times it was considered the elixir of life and was said "to make the heart merry", chasing away melancholy and gloomy thoughts. Lemon balm is a good fragrance to use after suffering emotional upsets and disappointment.

Bergamot is another orange-like fruit that produces a citrusy, uplifting fragrance. Its delicious scent is useful for lifting a dark mood. It combines well with lavender and geranium, two other uplifting fragrances. Lavender is a versatile fragrance because it can be used to influence mood in a variety of ways. This is because of its balancing action, which helps restore emotional equilibrium. Geranium is also a good balancer and helpful for stabilizing extreme mood swings, particularly those associated with hormonal imbalance.

Belonging to the same family as lemongrass and citronella, palmarosa is a scented grass whose fragrance resembles a cross between geranium and rose, with a slightly earthy undertone. Its uplifting scent has a stimulating action, helpful for easing feelings of listlessness and hopelessness, although rose is maybe a more obvious choice for inspiring blissful feelings. Whether a bowl of freshly picked blooms or a lighted incense stick, the scent of rose enlivens the heart and, like lemon balm, is particularly good for emotional healing. As with the citrusy scents, virtually all the floral fragrances are cheering and uplifting and it is really a question of choosing the ones that you like best.

For sharpness and depth, herbs or spices can be used alone or in incense blends. In the 16th century, Gerard, a well-known English herbalist, said that the smell of basil "taketh away sorrowfulness". Dill, a member of the carrot family, has a spicy aroma that is said to banish negative thoughts and bring contentment. It blends well with lemony scents. Nutmeg has a cheering scent that can be narcotic in large doses. For this reason, it is best used in pot pourris and herbal sachets rather than in vaporization. Allspice has an exotic, exhilarating aroma that resembles a blend of nutmeg, cinnamon, clove and black pepper.

left *The crisp, clean scent of these lemon and lime incense sticks (left) will lighten a dark and heavy mood. Lemons and limes grow in a hot, sunny climate. Their scent brings happiness and a reminder of summer days. The palmarosa sticks (right) have a refreshingly sweet, rosy scent. Also known as "Turkish geranium", palmarosa is cheering, lifting sadness and easing depression.*

Scented candles are available in all sorts of colours, shapes and sizes. These long, thin, orange candles look striking grouped together in a white bowl. Their piercingly sweet orange aroma smells exactly like peeled fruit. Orange is one of the most optimistic and cheerful scents and colours.

above *These handmade sandalwood incense cones capture perfectly the magic of this powerful scent, one of the oldest known perfume ingredients. Like a gentle breeze, the fragrance of sandalwood caresses the senses and is balm for a troubled soul.*

soothe

It's easy to get lost in a sea of troubled emotions. Fortunately there are many varieties of herbs, oils and incense blends to calm and restore. Combine them with delicate colours and soft pastel shades.

Feelings are part of life. Knowing how to cope with the difficult ones can be challenging, particularly in a hectic schedule. A gentle, soothing environment will help pacify the senses and calm inner tensions. Colours and design should harmonize, being quiet and understated rather than making loud attention-grabbing statements. Round shapes and flowing lines, rather than harsh angles and geometric patterns, are comforting, giving a subliminal message that evokes security and comfort. Scents too give subliminal signals, so choose ones to support how you feel. Soft lighting, relaxing background music and a little gentle fragrance wafting through the air is a soothing balm for jangled nerves.

Because of its complex chemistry, lavender is extremely versatile and has many uses. It is a soothing tonic for the nervous system, capable of calming stormy emotional states and rebalancing body and mind. The tiny, bluey-mauve dried flower-heads give a gentle fragrance to incense-burning mixtures. The delicate scent of rose is also calming to troubled emotions, particularly over-heated states such as anger and jealousy, or negative feelings of guilt and resentment.

Camomile is also known for its gentle, relaxing qualities. In ancient Egypt the plant was dedicated to the sun god Ra because it was considered effective in bringing down a fever. Camomile quietens heated emotional states, calms the nerves and eases stress-related tensions. Like lavender, camomile also has the great ability to restore equilibrium. Traditionally known as the plants' physician, camomile has been used to restore ailing plants back into life. A mild concoction of cold camomile tea can be used to water house and garden plants that look in need of a little loving care. As with lavender, the dried flower-heads of the camomile plant can be used in incense burning. Their balsamic, herbal scent suggests warmth and security, and combines well with a resinous base of mastic and myrrh.

above *A soothing cup of camomile tea makes a wonderful alternative to traditional tea or coffee. Camomile has a cooling effect, reducing fever and calming overwrought emotions, such as anger, jealousy or fear.*
below *The delicate floral scent of this lavender and rose sleep cushion will soothe troubled nerves. Both lavender and rose have been valued for centuries for their restorative powers.*

right *Vanilla is rapidly becoming one of our most popular scents. These vanilla-scented candles will create a calm and relaxed atmosphere and soften feelings of frustration and irritability.*

Traditionally, marjoram is said to comfort grief and calm angry or troubled emotions. It has a relaxing effect on the central nervous system and combines well with lavender, star anise and mastic in incense blends to combat tension and stress. Not to be confused with common sage, clary sage is another herb with soothing properties. The sweet, musky aroma has a euphoric quality that calms emotional distress, particularly states of fear, panic and anxiety.

Not surprisingly, there are many woody scents that can help with negative moods and inspire feelings of tranquillity. Standing next to a huge tree is calming in itself and can help put problems in perspective. It is hard to comprehend, but cedar trees at least 2,500 years old have been found high up in the forests of Lebanon, while cypress trees can live up to 2,000 years. Both of these majestic trees were sacred in ancient Mesopotamia and regarded as symbols of longevity and wisdom.

At that time, the word for the Lebanon cedar and frankincense was the same. Like frankincense, cedar slows and deepens the breathing, helping to restore inner calm. Cedarwood has a rich, warm fragrance that is harmonious and enduring. It is especially helpful during times of great anxiety. It is also useful for calming aggression and irritability, particularly when associated with work pressures. Sadly, there are very few cedar trees left in the ancient forests of Lebanon and most cedarwood today comes from its close cousin, the Atlas cedar.

Like cedar, the statuesque cypress has a long tradition of use in incense ritual. Its cones, leaves, twigs, resin and bark may be pulverized and burned alone or added to mixtures. Cypress has a strong, smoky, woody scent with a slightly spicy undertone. In times of change, cypress helps ease the transition from the old to the new, allaying fears of the future. It is particularly helpful for resolving long-standing grief and disappointments. The soft, sensuous scent of sandalwood is also useful for cutting ties with the past and for quietening anxiety, similarly the distinctive resinous scents of golden copal, frankincense and black storax. Black storax is best used in incense blends rather than by itself.

Whichever fragrance you choose, take the time to turn your incense burning into a leisurely activity that brings peace and calm. For a modern-day incense ritual, light a few candles and sit in a quiet spot where you will not be disturbed and let the fragrant smoke carry all your tensions and troubles away.

left *In Japan, star anise is honoured as a sacred tree. When dried, its yellow, star-shaped flowers turn a brownish, golden red colour. The sweet, aniseed-like fragrance has a balancing middle note. Sometimes known as the happy scent, star anise eases stress and worry, promotes calm and increases feelings of wellbeing. It is best used in small amounts as a complement to other aromas.*

Golden copal resin has a powerful effect on the emotions. Like a touch of tenderness, its fragrance when burned is soft and warming, a tonic for frazzled nerves. Black storax resin has a sweetly floral, balsamic fragrance that is calming and relaxing.

seduce

Spice up your love life and create a sensual seductive mood. In some languages, the word for "kiss" and "smell" is the same. Many scents are renowned aphrodisia and enjoying fragrance is sexy and fun.

Throughout antiquity, there has always been a strong association between scent and sensuality. Perfume is a traditional gift of love. In many sacred traditions, particularly in the East, making love was seen as a route to mystical experience and sensual pleasures were refined into a subtle art. Perhaps it is time to learn the forgotten language of fragrance and return to the tantalizing path of seduction. To set the mood for intimacy, accessorize with warm pinks, deep reds and oranges or rich maroons. Irresistible touch-me fabrics like suede, velvet, satin, soft lawn cotton or pure silk add a touch of luxury. For a romantic dinner, food and drinks with aphrodisiac properties, such as asparagus, shellfish, chocolate and champagne, should be included somewhere on the menu, while candles and soft music help set the scene for intimate exchange.

It's well established that like plants and animals, people also have their own natural smell. It's part of our biological makeup for attracting a mate. This is attributed to hormone-like chemicals known as pheromones, scent signals radiated by the skin. Certain animal and plant fragrances are similar to human pheromones and it comes as no surprise to learn that these are some of the most popular scents, used to stimulate sexual desire and the capacity for love and eroticism.

Musk, a natural secretion of the musk deer, is probably one of the most infamous, having an earthy, sensuous scent. In ancient China, courtesans were reputedly fed a diet flavoured with the substance so that their heated bodies would release its perfume in the passion of love. Today, the musk deer is a protected species, so other reputable sources have been found to produce a similar scent. The best substitute for musk comes from the seeds of rose mallow, a particular variety of hibiscus. Synthetic imitations are also widely available.

above *The intoxicating scent of jasmine liberates the senses. Erotic and sensual, it releases inhibitions and opens the doorway to love.*
below *Deep pinks and reds are some of the sexiest colours around. To set the mood for seduction, these holders can be used for powders, oils or candles.*

above *For a little fireside seduction, have a selection of incense powders ready for burning. In antiquity, a combination of agarwood, sandalwood, cinnamon and myrrh was the alleged favourite of kings, used to seduce and flatter the ladies at court.*

The erogenic aromas of both jasmine and rose make them practically synonymous with love and seduction. The delicate, star-shaped, white jasmine flower has an intoxicating fragrance that becomes more intense after sunset. The warm, sweet, exotic aroma has a musky, sensual undertone that men seem to find appealing. For centuries jasmine has been used in love potions and bridal garlands, its exhilarating aroma reputedly releasing inhibitions and liberating sexual fantasies. Jasmine is helpful for alleviating symptoms of frigidity or impotence in either sex.

Perhaps more than any other flower, the rose is a symbol of romantic longing, the desire for perfect partnership and mystical union. Traditionally associated with Venus, the goddess of love, the rose has the power to open the heart to intimacy. The most sweetly scented roses range from deep crimson red to a light blush-pink in colour and have a divine, distinctive fragrance that leads all the way to temptation. The modern-day custom of throwing paper rose-petal confetti at weddings pales into insignificance beside the Romans who made liberal use of the real thing, scattering fresh rose petals during feasts and celebrations and onto the bridal bed.

Ylang ylang is another intensely fragrant blossom customarily strewn on the bed of newly-weds. The ylang ylang tree is cultivated in Indonesia, its name translating as "flower of flowers" in the Malay language. The thick, waxy, yellow petals have an intensely sweet, voluptuous aroma that is soothing and enticing, an antidote to sexual indifference and emotional coldness. In common with ylang ylang, the fragrance of both neroli and patchouli has a narcotic effect, lulling inhibitions and taboos to sleep while secret desires unfold. Neroli is produced from the delicate white blossom of the bitter orange tree. At one time, the fragrance was a "signature" perfume of prostitutes in Madrid. More romantically, the blossoms are traditionally used in bridal bouquets to symbolize purity. In contrast, the bushy patchouli plant has a more earthy, erotic aroma which was very popular in the "free-love" era of the hippy sixties. The scent of patchouli penetrates deep into the emotions, dispelling indifference and calming sexual anxiety. It has a strong, long-lasting effect and is not to everyone's liking, so go cautiously if you are new to it.

Like patchouli, sandalwood has a long lingering fragrance. The antelope-brown-coloured powder is soft to the touch with a gentle, spicy undertone. Sandalwood's musky, exotic fragrance is profoundly seductive, caressing the senses into an intimacy that brings people together. Agarwood is little known in the West, but its effect is similarly magical, deeply relaxing and balancing for body and soul, the perfect love match.

left *This jasmine and rose pot-pourri is given an earthy eroticism with the addition of a little patchouli oil. A few drops sprinkled over the top of the dried flowers gives a deep, penetrating scent that lingers for hours.*

right *Traditionally known as the queen of flowers, the rose was first brought to Europe from Arabia, the land of the famous "Thousand and One Nights". In Arabia rose was a favourite perfume, along with musk, sandalwood and jasmine. The sublime fragrance of these incense sticks opens the heart to love and stimulates sexual desire.*

inspire

Everyone needs a little inspiration now and again. From humdrum tasks to creative projects, life is full of opportunities for original self-expression. Let the fragrance of swirling smoke inspire your dreams and stimulate your imagination.

People have always used incense to encourage dreaming and inspiration. This is because scent has a direct affinity with the intuitive part of the brain that governs creativity, imagination and emotional response. In contemporary society, with its emphasis on reason and logic, these qualities are often neglected and consequently are undeveloped in many people. Yet it is precisely these powers that enable us to problem-solve where logic fails and are frequently responsible for genius leaps of discovery, both large and small.

Some of the most inspiring fragrances come from some of the most traditional incense-burning substances. Frankincense has an ethereal fragrance that touches the emotions and opens the heart. Like the tree from which it comes, the resin is best suited to a warm, dry atmosphere. The full fragrance does not develop under cold temperatures and high humidity. Burning frankincense is said to improve the acoustics in a room, so it is a good fragrance to use when listening to music. If you prefer a lighter scent, eucalyptus has a fresh, herbaceous top note. Its fragrance is cooling and, like frankincense, helps deepen the breath. It is helpful for focusing concentration when engaged in creative thought.

The benzoin tree produces a resin with a delicious vanilla-like aroma. Most benzoin resin comes from Thailand or Sumatra. The Thai variety, usually referred to as benzoin Siam, has a lighter, more subtle fragrance than the one from Sumatra. Its soft, sensuous fragrance helps counteract apathy and stimulates the imagination. Benzoin combines well with sandalwood and cinnamon, particularly when involved in creative work such as painting or playing music. The combination helps lift mundane thoughts and aspirations to a higher plane. Benzoin is also grounding and has a fortifying effect on other fragrances in incense blends.

above and opposite *The refreshing scent of eucalyptus clears the mind and stimulates the imagination. It is like a breath of fresh air, reawakening the senses and inspiring creativity.* **below** *Cardamom is warming and uplifting. Its rich, sweet, gingery aroma improves the circulation and gets the creative juices flowing. New ideas are born with this scent.*

above left *These purple hyacinth blooms have an intensely sweet perfume that is both inspiring and energizing.*
above right *Writing is a good way of expressing your innermost thoughts. The scent of sandalwood, rose and cinnamon in these incense cones encourages self-expression.*

The warm woody aroma of sandalwood is delicate and uplifting. It helps loosen rigid attitudes and cynicism and encourages spontaneity and freedom of self-expression. It is a good scent to use when creativity is blocked and you are devoid of inspiration. One of the beauties of sandalwood is that it is a good mixer and works well with almost any fragrance, particularly with cinnamon. If you prefer a more intense floral scent, hyacinth refreshes the mind and improves concentration.

Well-known culinary spices such as cinnamon, cloves, star anise and cardamom are also helpful for stimulating creativity and intuition. Like frankincense, cinnamon opens the heart and creates a relaxing ambience. Cassia, or Chinese cinnamon, is less well-known but imparts a fuller, slightly more exotic aroma. Like benzoin, cloves too have a fortifying effect on incense-burning mixtures. The dried flower buds have a spicy aroma that is reputed to stimulate intuition, counter negative thoughts and encourage the recall of forgotten memories. Star anise is an evergreen tree that produces yellow, star-shaped fruits. When dried, their warm, sweet fragrance helps lift the spirits. Star anise is best used in small amounts with other fragrances. The aroma of cardamom is stimulating. It is a good scent to use when you are lacking in inspiration. Left whole, cardamom pods are useful in pot-pourri mixtures, combining well with orange and cinnamon.

Sacred to the Mayans, the copal tree produces three different types of resin: white, gold and black (or night) copal. When burned, each one has a strong effect on the emotions. White copal has a lemony, frankincense-like fragrance that is uplifting and exhilarating. The gentler golden copal is warm and soft, ideal for nurturing creative projects that are in their infancy. The Mayans burned it at sunrise to honour the power of the sun, a traditional symbol of creative energy. Black copal is grounding and mysterious. It is especially helpful for connecting with your intuition and psychic powers.

Another mystical plant of the Mayans was dream herb, or zacatechichi. The Mayans believed it encouraged lucid dreams, where it is possible to find answers to life's important questions. Today, scientists have discovered that zacatechichi strengthens dream phases during sleep, perhaps lending weight to what the Mayans believed. If you use your dreams as an aid to problem-solving, you may like to experiment with this plant. Burn a little of the herb before going to sleep, keep a notebook handy and write down any dreams as soon as you wake up.

right *Nature has always been a source of inspiration for artists. The heady scent of hyacinth stimulates right-brain, or creative, activity. It clears muddled thinking and inspires motivation.*

Open up the door to your **creativity** with fragrance. Aromas work on

the intuitive, irrational side of the brain that inspires creativity.

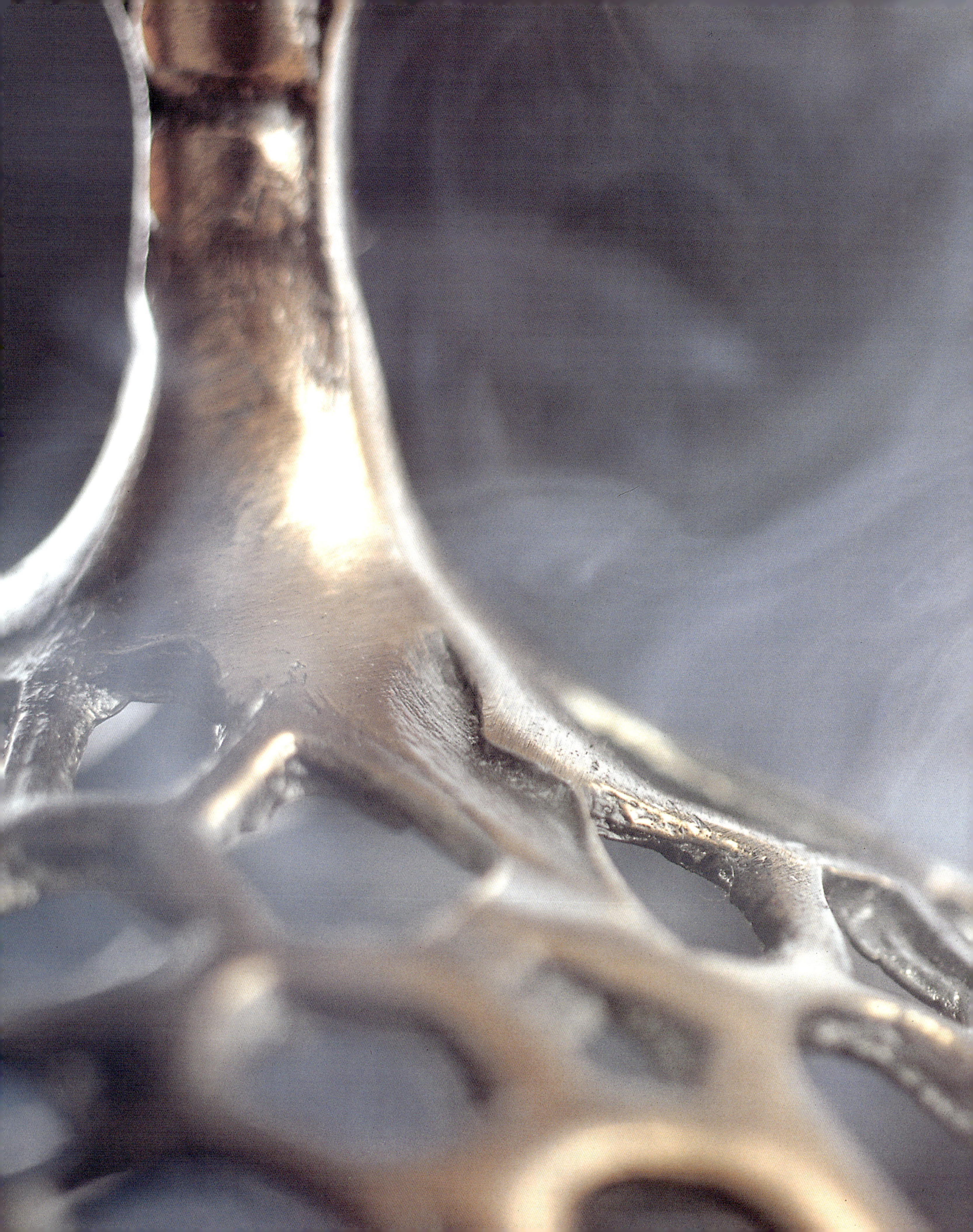

sacred scents

Traditionally aromatics were thought to represent the "soul of nature". From the very beginning, fragrance has always been central to ritual and sacred ceremony, used to cleanse and purify the atmosphere, heal sickness and to assist human strivings towards perfection.

cleanse

Incense has always been used to cleanse and purify, whether in sacred ceremony, medicine and healing, or everyday affairs. The fragrant smoke clears away germs and negative vibrations and encourages harmony and wellbeing.

Aside from the dirt we can see and touch, invisible "psychic" dirt can pollute the atmosphere. It is as though our moods, thoughts and actions leave an energetic "imprint" behind us. For instance, we speak of an atmosphere being charged with tension or thick with fear. This psychic dirt can also "stick" to people and their possessions. Fragrant incense smoke is a simple yet powerful way of clearing the air after an argument or bouts of sickness, or even to neutralize the atmosphere after a round of social visits. It's also a good idea to "space clear" your environment when moving into a new home. This removes any lingering vibrations from the previous occupants and helps establish your claim to the property.

There are many purifying scents. In the Native American tradition, sweet grass, desert mugwort and sage are used in cleansing ceremonies. Sweet grass has a pleasant fragrance, reminiscent of freshly mown grass, that is said to attract positive vibrations. Desert mugwort is a cleanser and protector against negative influences. It is typically used to cleanse precious items by suspending them in the smoke. White sage is one of the best fragrances for clearing negative vibrations and restoring harmony, although other varieties of sage have a similar effect. In a practice known as "smudging" dried sage is used alone or combined with other herbs and tied together in "smudge" bundles or sticks. The bundles are lit and made to smoulder, the smoke clearing the air and the ashes used to "smudge" over a surface to protect against negativity. To clean your aura, fan the smoke and then draw it towards yourself, taking it up over your head and down towards your feet. This traditional Native American practice is becoming more widespread and herb bundles are available at specialist retail outlets. Some bundles also contain dried lavender heads or cedar leaves, both of which have purifying properties.

above and below *White sage has a long tradition of use in cleansing and purification rituals. Holding an object in the smoke is believed to clear away old energy patterns and to make it as good as new.*

opposite *Every now and again cleanse your jewellery with incense smoke. Frankincense is one of the most powerful cleansers. In some parts of the world it is used to freshen clothes.*

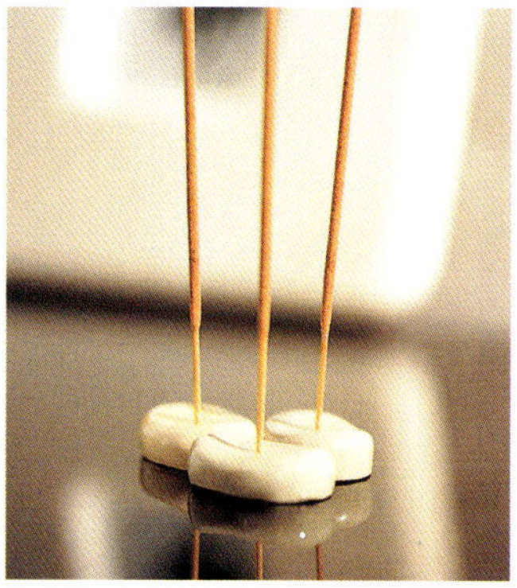

right and far right *The scent of frankincense is given a twist with the freshness of orange and lemon in these blended sticks. All three substances are antiseptics. Use this combination to clear the air after an argument. Frankincense and lemon are purifying, while orange will help restore cheerful feelings and harmony.*

Frankincense is another powerful aroma for cleansing and clearing the atmosphere. It is also a potent antiseptic. It is still used for daily hygiene in parts of Africa and the Middle East. Of course, frankincense is more commonly known for its associations with sacred ceremony. Its unique fragrance purifies the atmosphere and the thoughts of all those present in preparation for prayer or meditation. At home, frankincense is a good incense to use when you want to clean personal items that have a long history or are in frequent use, such as antiques or pieces of jewellery. It will help the object regain its original freshness. Frankincense is also useful when the atmosphere seems thick and heavy with trouble and unease. Other useful incenses for clearing a heavy atmosphere include copal, cedar, camphor, sandalwood, lemon and elemi.

In ancient times, hyssop was a holy herb, valued for its purifying properties. It has a spicy, aromatic scent that is similar to thyme, another strong antiseptic. Brooms made from hyssop were used to clean out temples and sacred places and later it was a popular strewing herb. During the plague, aromatic plants such as hyssop, juniper and pine were burned on bonfires to keep the air germ-free. Until relatively recently, fumigation was a standard practice. In European folklore juniper was believed to offer protection against evil influences. In the 17th century, wealthy households hired servants to keep the house smelling sweet and free from germs, while during the First World War, sprigs of dried juniper were burned in French hospitals to disinfect the atmosphere. The plant's evergreen tips as well as its dark blue berries give a fresh, bittersweet fragrance that is a powerful disinfectant.

Finally, eucalyptus is one of the most powerful antiseptics. In Australia, for example, eucalyptus trees are traditionally grown in swampy areas to help prevent the spread of malaria. This makes eucalyptus invaluable during illness or in the winter to keep colds and flu at bay. Its penetrating menthol aroma blends well with hyssop or pine.

left *The botanical name for sage –* Salvia *– comes from the Latin meaning to save or cure. Sage is one of nature's most potent antiseptics and a traditional symbol of wisdom. It has been valued since antiquity for its cleansing and healing powers. Its warm, fresh, camphorous aroma is penetrating and uplifting. The fresh herb can be brewed and made into a tea.*

heal

According to an ancient Chinese saying, "a perfume is always a medicine". Incense and aromatics have a long and varied tradition of use in medicine and healing. They can be used to alleviate many symptoms of distress.

Medical papyri from ancient Egypt reveal how sick people, or at least the painful parts of their bodies, were exposed to fragrant smoke. This type of incense burning is one of the oldest reported healing practices, still used in Ayurvedic medicine and by many indigenous peoples around the world. Because many aromas have relaxing and soothing qualities, they are helpful for easing pain and muscle tension, for treating respiratory illnesses and insomnia and, of course, for restoring emotional equilibrium. For everyday self-help you can experiment with incense and aromatics, although you should not treat serious problems yourself.

Frankincense and myrrh are two of the oldest incense substances, and have many healing properties. Frankincense aids cellular renewal and helps to heal wounds, both physical and emotional. To heal a rift in a relationship, take a photograph of the person and light some frankincense in front of it, sending the person thoughts of love and forgiveness. Burning frankincense is also said to improve the circulation, lower fever and to reduce toothache and rheumatic pain. Because frankincense has a calming, deepening effect on breathing it is also a traditional remedy for opening up the airways and alleviating bronchial conditions.

Myrrh is also calming and promotes healing on many levels. In the ancient world it was widely used for pain relief and in the treatment of wounds. Because myrrh reduces inflammation and encourages the skin to heal, it is useful for common skin problems such as chapped skin, acne and eczema. In Asia, women traditionally bathed in the smoke of myrrh resin for beautiful skin. On an emotional level, myrrh is useful for calming agitated states of worry and fear, bringing peace and inner calm. Because of its cooling properties, myrrh is also a good resin to use after an argument or when angry. In Arabic countries it is used as a tonic for stomach tension.

above *More than 4,000 years ago, the Chinese were using cinnamon to strengthen and heal the body. The scent of cinnamon is warming and comforting. It combines well with cloves, a natural antiseptic.*

below *The fragrance of rose is a gentle panacea for emotional upsets. Rose helps to restore self-confidence after disappointments in love and eases emotional pain.*

above *In the ancient world, myrrh was used in healing salves and lotions. Burn myrrh as incense when emotions are running high. Its musky, earthy aroma drifts through the air, bringing happiness and restoring harmony. Myrrh combines well with rose.*

above *In the past, juniper was revered as holy, the tree of life that protects against evil. Today, we associate juniper with cleansing, healing and strengthening body, mind and soul. The berries have a fresh aroma when burned.*

right Scented pine cones are another way of using fragrance for wellbeing. These cones have been soaked in a mixture containing sandalwood, cinnamon, orange and cloves.

Sandalwood is an immunity booster and provides protection during times of vulnerability. Because of its soothing, anti-inflammatory properties, it helps with stress-related conditions, such as insomnia and irritability, as well as easing muscular pain and tension headaches. It is also useful for turbulent emotional states.

Calmus and storax were also widely used in the ancient world. The tangy, cinnamon-like fragrance of calmus root is said to strengthen the nervous system and enhance worldly success. It is helpful during times of emotional upheaval and when plagued by self-doubt and low self-esteem. It is very strong when used by itself, but mixes well with storax and mastic. If it is not available, cinnamon has a similar action. Storax is calming and relaxing for the nervous system. It is useful for treating insomnia and stress-related symptoms. It works best when combined with other fragrances such as mastic, myrrh, cloves or cinnamon.

The aroma of mastic resin is fresh, clear and lemony. In northern Africa, it is used as a tonic for exhaustion and to speed up recovery. Creative visualization is believed to increase the plant's healing powers. In times of debility, imagine a protective shield of light encircling the person in need of healing – this could be yourself or someone else. This shield keeps away negative influences yet allows positive energies to filter through. Repeat the exercise each evening and again in the morning.

For joint pain, muscular aches and general debility, juniper and coriander are two useful plants. In the Middle East, coriander seeds are burned to treat stress, anxiety and tension headaches and to lift depression. Juniper is a potent detoxifier and helpful for arthritic conditions and skin problems, as well as flushing out unwanted emotions. Both coriander and juniper protect against negative energies. In areas where there is a lot of coming and going, such as the hallway or kitchen, they will help stabilize the atmosphere. They are also useful to neutralize and balance the energies in a new home. Coriander helps to create a harmonious atmosphere, encouraging people to communicate and make peace with one another. It combines well with frankincense, mastic and myrrh.

There are, of course, many flowers that bring comfort, including rose, lavender and freesia. Rose helps a broken heart to mend, while on a physical level, it regulates the action of the heart, improving the circulation and toning the blood vessels. It is especially helpful for women's health, having a tonic effect on the reproductive organs and helping balance the hormones. Lavender calms stormy tempers and extremes of emotion. It has a steadying influence that helps remove emotional conflict. Lavender is also good for muscular aches and pains, fevers, infections and stress-related problems. The scent of freesias is uplifting and will cheer you up when recovering from illness.

right In the language of flowers, freesias bring optimism and hope for the future. They are often used in bridal bouquets as a symbol of the promise to love one another. Freesias have a sweetly penetrating, distinctive scent that brings healing and harmony.

transform

The oldest and perhaps the most profound use of incense is in religious ritual and sacred ceremony. As solid matter transformed into scented smoke, a link was established between the material world on earth and the spiritual realm in the heavens above.

To the ancients, fragrance was nectar for the gods and a sweet scent indicated the presence of angels. The aromatic spirals of smoke were a flight-path for prayers or else a way to self-realization and enlightenment. Since the earliest times, specific fragrances have been used to awaken and strengthen spiritual energies, to connect our everyday life here on earth with the vastness of the cosmos. One of the beauties of using incense is that it does not matter whether you subscribe to any particular spiritual path. Almost everyone appreciates some quiet time for reflection. To cultivate inner peace, burning incense refreshes and nourishes the soul. Use it to transform your life into a spiritual experience.

There are many fragrances to choose from, including dammar (cat's eye resin) to illuminate darkness, cedar for wisdom or elemi for new beginnings. You can also look out for blends such as Nag Champa from India or the hand-rolled Tibetan sticks used in Buddhist monasteries. Some of the most precious incenses for ritual are frankincense, copal, myrrh and sandalwood, all with a long tradition of use.

The fragrance of frankincense is spiritually uplifting and deepens mystical experience. Myrrh has a more grounding, earthy quality and brings awareness to the spiritual dimension in the material world. Myrrh is traditionally associated with feminine spiritual power.

Believed to be a gift from the gods, the most precious copal resin came from trees that were hit by lightning. To the Mayans, copal was so sacred that it could only be handled with special instruments by the priest. The incense was not so much a pathway for prayer, but a prayer in itself. Burn copal for flashes of insight and self-awareness.

With more than 4,000 years of recorded use, sandalwood is probably the oldest known perfume in history. In India it is used to build temples and by yogis as an aid to meditation. Sandalwood has a unique aroma that quietens the mind and stills the soul. It creates a feeling of inner peace and being at one with the world.

above *The delicate fragrance of these sandalwood cones retunes the mind from a state of anxiety to one of calm tranquillity.*

below and opposite *Create a special place for incense burning and transform it into a spiritual experience. The holy smoke will raise your thoughts from the level of the mundane to the sacred. Here, natural objects serve as a reminder of our connection to nature.*

fragrant ritual

Rituals are a way of marking important occasions. They are also a way of creating sacred space in daily life and almost any activity can be elevated from the ordinary to something special. Incense burning may be performed as a ritual act.

Rituals are a way of honouring transitional states as we move from one stage of life to another. They provide an opportunity to step outside routine ways of thinking and behaving in order to connect with life's bigger picture, symbolized for some people as a divinity or the "higher self". Rituals are a time to reflect, to give thanks or to ask for guidance and support. They may also be a way of celebrating and having fun. Traditionally, major life events, such as birth, marriage and death, are acknowledged with some kind of sacred ceremony. Similarly, religious and seasonal festivals are also acknowledged by ritual acts.

However, rituals need not be reserved for momentous events, but can be a way of bringing spirituality into everyday life. Whether eating, taking a bath or making love, the experience can be transformed through ritual. Burning incense can help you achieve this. Neither solid nor disembodied, the fragrant smoke is a powerful symbol of transition, of crossing the boundaries between one space and another.

Traditionally, an altar or shrine is used as a focus for ritual activity. It is not necessary to have any religious belief to create your own altar. It means setting a space aside to symbolize those things that have meaning for you. Altars can be made anywhere; they can be indoors or outside – it really does not matter. The point is to create something which inspires and nurtures your soul. This is very much an individual choice. You may like to use some kind of religious icon or object, such as a Buddha statue, a cross or prayer beads. Alternatively, secular images such as family photographs or a precious artefact are suitable. Candles and incense will help create a sacred focus. Lighting them at the start of a ritual symbolizes opening up to the "light", the higher spiritual powers which may be called upon for healing, prayer or celebration. Choose the colours and scents accordingly.

opposite *An altar is a sacred space. Select objects that resonate with your soul to go on your altar. Here, jasmine flowers and a Buddha statue are used in ritual incense burning.*

above and below *The Japanese distinguish between incense burning for pleasure and burning incense to the Buddha. Candlelight and flowers represent the innocence of the heart.*

Koh-doh is a light-hearted way to enjoy incense and to extend your scent repertoire. A koh-doh cup is filled with white rice ash, and a piece of bamboo charcoal is pushed down into the cup (as shown). The charcoal should then be pushed down into the ash. After making a "chimney" over the charcoal, place the mica on top, followed by the ground incense. Agarwood (back) is the primary ingredient in koh-doh. Although it may not look very special, its smell when burned is out of this world. Sandalwood chips (front) are also typically used in koh-doh.

The Japanese enjoy a sophisticated fragrance culture. Koh-doh,

above *The finest quality bamboo charcoal comes from Japan. Unlike other charcoal blocks, it does not contain saltpetre and is available in a range of sizes, shapes and even colours. The silver coated blocks (left) have the advantage of being clean to handle. The ones with the red dot (right) are the traditional type used in koh-doh ceremony. They also look very stylish.*

Perhaps more than any other nation, it was the Japanese who developed the ritual of incense burning into a fine art. The word for incense in Japanese (*mon-koh*) translates as "listening to incense". To appreciate incense, you must be attentive to the process of lighting and burning it, noticing how the fragrance affects you in subtle ways. To do this is to bring yourself into the here-and-now so completely that there is no room for thought of past or future. Only then are you able to "hear" the smoke as it carries you away to a state of complete serenity.

Such a level of sophistication is evident in koh-doh, the "way of fragrance", developed in the 14th century. The Shogun Ashikaga Yoshimasa demanded the classification of all incense-burning substances and established rules that regulated their use. Social games for incense appreciation were developed with a master of ceremonies presiding over the event. The idea was for participants to "listen" to the fragrance and identify it correctly. Experiencing a fragrance consciously during koh-doh was said to expand the soul, stimulate creativity, and delight the senses.

Today koh-doh is undergoing a renaissance in Japan and is rapidly becoming a popular after-dinner activity in the West. It is possible to purchase a complete set of koh-doh accessories from specialist suppliers, but you can also improvise. You need an incense-burning cup, white rice ash (or sand), a flat knife, a metal chopstick or skewer, bamboo charcoal, a pair of tweezers, matches, a piece of mica and the incense ingredients.

Fill the bowl two-thirds full with the ash and press it down with the knife. Light the charcoal and, using the metal tweezers, place it in the centre of the cup and push it down into the ash so that a mound forms over it. Make an air hole in the ash to form a "chimney" over the charcoal with the metal chopstick. Place the mica on top, and then the incense on the mica "hot plate". The bowl is passed in a clockwise direction with each participant lifting the bowl to their nose and inhaling the fragrance, turning their head away from the bowl when breathing out. After inhaling two or three times, the bowl is passed on to the next person and the person writes down what they guess the fragrance to be. Notes are compared at the end. The fragrance may also be used to stimulate story-telling sequences.

In the beginning it is best to limit the number of incenses used in koh-doh to five or six, with agarwood (*jin-koh*) being the essential item. Other traditional koh-doh incenses include sandalwood, Borneo camphor, frankincense, myrrh or patchouli, plus some spices such as star anise, cinnamon and cloves.

the way of incense, involves the burning of fine woods and resins.

seasonal scents

The seasons are nature's moods, each with its own accent. Spring is fresh beginnings, summer is abundance, autumn is decline and winter a time of rest. Use fragrance to harmonize with seasonal change and be in tune with the natural rhythms of life.

spring

After the long dark days of winter, the first green shoots of new life appear. Spring is a sign that life goes on. It's a time of birth and growth, of hope and optimism for the future. Fragrances are clear, fresh and intense.

Spring's new beginnings are heralded in nature by bursts of colour and new life. Lively yellows, verdant greens and shots of bright blue, purple, pink and white adorn gardens, parklands, woods and hedgerows. Some of nature's sweetest and most fragrant perfumes scent the air. Early-flowering bulbs such as lily-of-the-valley, jonquil, narcissi and hyacinth are strongly aromatic, while a little later many trees, shrubs and climbers produce fragrant blossom. These include the delicate scent of apple, the intoxicating fragrance of lilac and the heavenly smells of magnolia, wisteria and yellow azalea. White Easter lilies also have an exotic and haunting scent. Spring is nature's celebration of life and a good time to refresh and reawaken your senses. Now is the best time to give your life a complete overhaul.

The most obvious place to start is indoors. Traditionally, spring-cleaning was an elaborate ritual of cleaning the house from top to bottom, clearing away winter's dust and dirt to prepare for new beginnings. Of course, cleaning the house is something that happens all year round. However, it is still a good idea to set aside some time and do a "deep clean" at least once a year. Use fresh citrus fragrances in your cleaning products and remember to clean under beds, behind cupboards and in any dark corners that generally get overlooked. Hire or purchase a carpet cleaner to freshen up carpets, rugs and upholstery and change winter bedding to something more lightweight. Pack away winter clothes, adding incense sachets or scented cones to keep them smelling fresh and to keep moths away. Useful scents include lavender, thyme, amber, sandalwood, lemongrass, cinnamon, pine and peppermint.

As you clean and tidy away, take stock of all your possessions. Review your wardrobe and throw out any clothes that you no longer wear. Similarly, go over your collection of books, magazines and CDs and get rid of any you do not really need; also any knick-knacks, children's toys and games and anything else that you're holding on to "just in case". Be ruthless – it's worth it. Letting go of clutter frees up space in your life for the fresh and new.

opposite and below *The liquid gold of amber resin looks almost good enough to eat. Amber is fossilized pine resin, millions of years old. It has a fresh, piney-camphorous scent. In spring, the sun's powers are increasing. In ancient Greece amber was known as "sun stone" and it was associated with renewal and alertness.*

Clearing clutter can have a tremendous psychological impact. It is like letting go of part of yourself each time you throw something out and you may experience some resistance. There are certain fragrances that can help you. Juniper has a detoxifying effect on body, mind and soul. It is a good aroma to use for mental and emotional clarity, helping you decide what to keep and what to throw away. It is strengthening and healing. Similarly, eucalyptus can stop you daydreaming and help you focus on the task in hand. Its aroma creates a feeling of space and freedom. Fresh forest scents such as pine, cedar and cypress help to strengthen the nerves and give support. Pine is good for clearing negative emotions and unwanted psychic energies. Cedar relaxes tension and cypress helps ease the transition from old to new.

At first a clean slate may be a little awe-inspiring and it is tempting to fill up all the new space with the familiar. However, this is a good time of year to experiment with new ideas and ways of being. Spring is about breaking away from routine and putting into action any plans or ideas that you have had on the back burner. This could mean signing up for a course, looking for a new job or property, or making plans to start or extend a family. If you are simply feeling restless and needing a change, altering the colour of your front door or interior decor can do the trick. Similarly, you may want to experiment with moving furniture around and redesigning a room's layout. Do not rush into anything, but carefully plan the effect you want to achieve and keep your eyes open for inspiration as you are out and about.

Cheering, uplifting fragrances are especially useful if you wish to stimulate new beginnings and creativity. After the darkness of winter, all the citrus scents will help get your creative juices flowing. Lemon and lime are revitalizing. Grapefruit is especially helpful for procrastination. It is also a good confidence booster and eases any tension or anxiety about doing something new. The sharp, keen scent of bergamot is good for frazzled nerves and the feelings of insecurity that accompany new ventures. Orange is bright and lively and creates a positive outlook, while petitgrain is invigorating and inspiring. Alternatively, the lively lemony scent of lemon balm is cleansing and revitalizing for heart and soul. Use it alone or with light florals such as lavender, camomile or rose geranium.

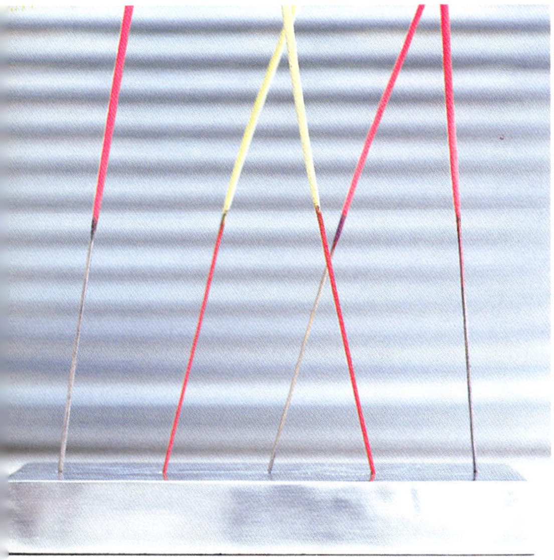

Fresh green herbal scents are also stimulating and can help shake off feelings of sluggishness, indecision and lack of motivation. These include coriander, basil, sweet grass and rosemary. Or, for a spirit of hope and new beginnings, burn a little elemi resin. Its spicy lemony scent combines well with lemongrass or other citrus scents.

left *The strong, bright colours of these incense sticks are perfectly in tune with the spring season. The yellow sticks are lemongrass, the pink ones rose geranium.*

right *Spring breeze is the name given to these turquoise incense cones. The fresh, penetrating scent of eucalyptus combines with lavender and camomile.*

summer

Long hot sunny days spell summer, the season of relaxed and easy living. Summer is nature's bounty, a gift to be enjoyed. It is a time to recharge and savour the sweetness of life. Scents are fruity, green and floral.

Summer is a season of delight and exuberance. Everywhere in nature a riot of colour and fragrance assaults the senses. These aromas include wallflowers, roses, sweet peas, phlox, pinks, lavender and honeysuckle. In late summer, flowering plants such as stocks and nicotiana give off their heady perfume in the evening. Summer berries and other fruits as well as herbs are also strongly aromatic. The colours of summer are hot pinks, crimson reds, royal purples, vibrant yellows and warm oranges, tempered by cool whites and creams and washed-out blues, pinks and mauves.

Confronted by such abundance, summer really is about unwinding and enjoying life to the full. This is the season for holidays, days out and generally having fun. More activities can take place outdoors, with plenty of scope for barbecues, picnics and summer parties. At first it may seem strange to think of using incense outdoors, but it can be used to good effect to keep insects at bay, as well as adding another subtle nuance of fragrance on a warm summer's evening.

Citronella, vetiver and lemongrass are botanically related and universally disliked by insects. These aromatic grasses grow in Sri Lanka and other tropical areas. Citronella is widely cultivated for commercial use in insect-repellent preparations. It has a fresh lemony smell that is quite distinctive. Citronella is available as an essential oil and in incense spirals. In India vetiver root is traditionally woven into awnings and blinds to help keep flies and insects away. Vetiver has a smoky and earthy aroma, slightly reminiscent of myrrh but with a lemony overtone. It can be used in small amounts in incense blends and combines well with sandalwood, jasmine, cedarwood or lavender. As its name suggests, lemongrass has a strong lemony scent. It is another useful aroma for keeping insects away and may be combined with lavender. Other useful insect deterrents are lavender, eucalyptus and tea tree.

opposite and below *The ice-cream colours of these summer roses look cool and fresh against a backdrop of crisp white. The tuberose incense sticks complement the fragrance of rose. Tuberose is one of nature's most exotic blooms.*

When packing your holiday bags, make sure you include a variety of different incenses and essential oils. Using your favourite fragrances in your hotel room helps to personalize the space and make it yours. There are also several aromas that can help jet lag. To encourage your system to relax, cypress, petitgrain and lavender go well together. Other calming scents are camomile, neroli and cypress. Alternatively, for a morning wake-up call, a combination of peppermint, geranium and rosemary is enlivening. Other refreshing morning-time scents include bergamot, grapefruit, lemon and rosemary.

Lazy summer evenings are perfect for romance. Whether you already have a partner or would like to meet someone new, certain fragrances can conjure up the mood for love and spread a little romance in the air. Many of these voluptuous and exotic scents originate in hot climates. Tuberose is native to Central America. Related to narcissus, the plant has sublimely fragrant, lily-white flowers. It blends well with other intense floral scents such as gardenia, rose, jasmine, neroli and ylang ylang, all of which are renowned aphrodisiacs. Similar to ylang ylang, the cananga tree is native to tropical Asia. It flowers all year round and bears large yellow blossoms with a tenacious and heady aroma. Cananga may be mixed with copaiba balsam, labdanum, jasmine or neroli.

Sometimes pressures of work and other factors can turn summer into a slog. If your summer is far from idyllic, then celebrate the joy of the season with appetizing and aromatic summer food. Add herbs such as fennel, dill, mint, chives and parsley to a handful of freshly picked lettuce, rocket and baby spinach for a sensational green salad. A few edible flower-heads such as nasturtium, borage or marigold will add vibrant colour. Soft summer fruits smell delectable and are packed with goodness. Strawberries, raspberries, blueberries, blackcurrants and redcurrants as well as all varieties of melon are good to eat at any time of the day. Similarly, luscious peaches and nectarines, juicy plums, apricots and mangoes smell delicious and are honey-sweet.

In some parts of the world, summer's long, hot days and short, balmy nights are celebrated at the summer solstice. This is the time of the year when the sun is at its zenith, bringing warmth and light so that crops may ripen. It is when creative energies are at their strongest. The ritual burning of incense can help you connect with this high point of the year and its special energies. Elecampane (inula) root was traditionally used by the Celts in their summer solstice festivities. The plant was believed to symbolize the energy of the sun and many healing properties were attributed to it. It can be combined with frankincense, mugwort and myrrh for a ritual incense blend to bring you in touch with the sun's magical power.

above and opposite *With so many fresh flowers available at this time, there is no shortage to choose from when it comes to indoor arrangements. Roses and freesias are both intensely fragrant. When combined they produce an aroma that is rich and intoxicating.*
left *These scented beads from Japan are an unusual and versatile way of enjoying fragrance. They can be used to scent linen and clothes, added to pot pourri or even worn next to the body. Entitled "golden nectar", their spicy floral scent is reminiscent of lazy summer days and warm, balmy tropical nights.*

autumn

This is the time of early morning mists and days that grow cooler as the earth moves away from the sun. Autumn is also a time of harvest and decline, as resources are garnered and stored for the long winter. Fragrances are woody and mellow.

After the glorious crescendo of summer, in autumn everything in nature begins to contract. Russets, golds, amber, dark reds and nut-browns are the hallmarks of the season as leaves change colour and begin to fall, blown about by the blustery autumn winds. The smell of wood smoke, rotting vegetation and ripe fruits mingles with late-flowering roses and damp grass. Autumn is a rich tapestry of colour and fragrance, a time to reap nature's bounty as the old withers and dies. It is a season of endings and completion, of fullness and satisfaction, traditionally acknowledged by rituals of gratitude and celebration. Autumn is a turning point between the free and easy living of summer and the more sombre constraints of winter. It is a time to remember the past, while preparing for the days ahead.

Surely one of the most spectacular sights of autumn is the harvest. Foods appear in a glut at this time of the year to tempt our taste buds. Grains, nuts and seeds plus a colourful medley of fresh fruits and vegetables are in abundance, while many thanksgiving festivals and feasts occur in this season. As the nights draw in, autumn is also a good time for entertaining, a time when you can invite friends to informal suppers or bonfire parties.

You can also use scents to celebrate the season. Earthy, woodsy fragrances represent autumn. Cedar has a rich, warm aroma that creates a comfortable, refreshing atmosphere. Used by itself, it creates a lot of smoke, so is best used outside on the bonfire or a little added to incense mixtures in powder form. It combines well with frankincense, jasmine, geranium and eucalyptus. Alternatively, the earthy smell of patchouli is perfect for autumn fireside evenings. It has a heady aroma that is said to resemble damp jungle undergrowth. From the tropical rainforests, rosewood has a sweet woody fragrance with a floral undertone. On grey and depressing autumn days, its scent will cheer you up. Scotch pine is also uplifting. Its refreshing scent is a reminder of evergreen forests, fresh air and the magic of nature. Patchouli, pine and rosewood blend well together.

above *The woody, earthy scent of these fruit-shaped scented candles is wholly in tune with the season.*

opposite and below *Autumn is a good time of year to look out for unusual plant pieces for pot pourri. Almost any natural object can be included in pot pourri. Spices such as cinnamon, cloves, nutmeg and allspice make a good base for woody, earthy or fruity scents.*

left *Autumn's nostalgic mood often brings old memories to the surface. Burning sandalwood can help you let go of the past and free yourself to move forwards. Its soothing aroma is uplifting and helps you to find hidden inner strengths.*

Golden autumn days are an invitation to venture outdoors. Look out for plant pieces for making pot pourri, incense or herb sachets. Angelica seed-heads, lavender spikes, juniper twigs, pine needles and cones, rosemary leaves and elecampane (inula) root are all useful ingredients. Similarly, the fallen branches of fragrant woods such as apple, cherry, olive or laurel are worth picking up when you find them. Let the ingredients thoroughly dry out before using them in recipes or for burning.

Autumn is a good time to take stock of your life and to trim your sails, letting go of any excess baggage you have accumulated throughout the year. This may be physical or emotional, for autumn is a time of memories and coming to terms with the past. In this "soul-clearing" process, buried feelings often come to the surface. Light a candle and burn some incense to help you let go of any outdated attachments. Gold, orange or yellow candles are perfect for autumn, while there are many suitable aromas that you can use, particularly frankincense, myrrh and sandalwood, and also agarwood.

Brighten up autumn's cool, damp days with spicy pot-pourri mixes. Or, be in tune with woody, earthy scents.

Frankincense works at a deeply spiritual level. Its distinctive balsamic aroma reaches the darkest corners of the psyche to heal old emotional scars, restore faith and encourage an optimistic outlook. The more earthy fragrance of myrrh also helps to heal the past. This golden amber resin with its smoky, musty aroma is perfectly in tune with the autumnal mood. Like frankincense, myrrh instils feelings of inner peace and calm. Sandalwood is also calming and penetrates deep into the emotional world. In India, it is traditionally burned at funerals, as its aroma is believed to liberate the soul.

Little known in the West, the magic of agarwood has been kept a closely guarded secret in Middle and Far Eastern countries for thousands of years. Its indescribable fragrance needs to be experienced to understand how special it is. It has been compared with a blend of sandalwood and ambergris. It is said to open up the heart to compassion and forgiveness. Pieces of the wood are extremely expensive, but you only need burn a small sliver at a time.

right *As the weather gets cooler and the nights draw in, lighting a fire creates a mellow and inviting atmosphere, perfect for entertaining or enjoying relaxing autumn evenings. There are many suitable incenses for burning at this time of the year, but patchouli, pine, rosewood or cedar all seem to hit the right note. Rosewood is particularly comforting when you are finding it difficult to adjust to the change of season.*

winter

Crisp icy weather and clear open skies reflect the radiance of winter. The days are short and nights are long as nature slumbers in preparation for the activity of spring. Fragrances are warm and comforting, spicy and invigorating.

Stark silhouettes stand out against the winter skyline, the skeletons of trees picked bare after autumn has finished its work. Winter is uncompromising; the inhospitable landscape demands strength and endurance in order to survive. When food and shelter is at a premium, hardy evergreens buffer the elements and offer sustenance to wildlife. Scents in nature are much harder to find, but certain varieties of viburnum or honeysuckle (*Lonicera*) are fragrant. *L. fragrantissima* is an evergreen shrub with creamy white flowers, while the scented viburnum has rose-pink (*Viburnum grandiflorum*) or white (*V. fragrans*) blossoms. Indoors, houseplants such as jasmine or gardenia come into bloom in winter. Both have fragrant white flowers. Winter's colours are luminous and striking: ice blues, slate greys and frosty whites combine with dark browns and greens and flashes of crimson, orange and purple.

Winter is the indoor season. It is the time to close the shutters and to spend more time alone or with close friends and family. The focus is on cultivating harmony, while nurturing body and soul. Wholesome food, plenty of sleep, brisk walks and generally taking care of your needs is required. A mellow and inviting ambience can be created with warm, spicy scents and rich pungent aromas. Many of these are familiar culinary ingredients. For winter evenings, experiment with nutmeg, cloves or cinnamon. The evocative fragrance of nutmeg is warming and stimulating. The Chinese believe it restores the body's energy, but it should be used cautiously as too much can over-excite the nervous system. It is probably best used in pot pourri or blends rather than by itself. It combines well with both orange and cloves. Cloves have a spicy, sweet aroma that is a powerful antiseptic and a useful protection against seasonal epidemics. Using cloves intensifies the fragrance of incense-burning mixtures. Cinnamon strengthens the immune system and protects against colds and flu. At one time more valuable than gold, cinnamon is one of the most nourishing and potent winter fragrances.

above *Star anise and cloves have been used as fillers in this unusual herb sachet. The fine see-through fabric retains the visual interest of the spices as their fragrance is released.*

opposite and below *Slices of orange and sticks of cinnamon form the basis of this pot pourri. Both scents are traditionally associated with the winter season. Orange is bright and cheering, cinnamon warm and nurturing.*

above and opposite *Winter's colours are sharp and clear. These red and blue incense sticks have an uplifting, camphorous aroma.* **above right** *Burn these colourful cones for some light-hearted fun during the dark days of winter. They have names such as orange vanilla, mulberry wine and chocolate dream.*

Traditionally, mid-winter has been marked with feasts and celebrations, both sacred and secular. The Romans enjoyed unrestrained merry-making during the feast of Saturnalia. For some modern-day winter cheer, orange is a reminder of the joy and frivolity of summer. Together with cinnamon, oranges are one of the traditional ingredients in mulled wine. Studded with cloves, they make a fragrant table decoration in winter.

Winter is also the chocolate season, the time of year when a little decadent self-indulgence seems justified. Hundreds of years ago, the Aztecs were flavouring their chocolate drinks with vanilla. Today, chocolate and vanilla are used in an endless variety of mouth-watering delicacies that are some of our most popular and delicious treats. The fondant aroma of vanilla is one of nature's sweetest scents and the smell of chocolate seems heaven-scent. Both fragrances are warming, ideal for snuggling up to on cold nights. Vanilla may also be enjoyed as a room fragrance, in scented candles or incense. As an alternative, the sweet balsamic fragrance of benzoin resin is similar to vanilla. Its sensuous warmth mixes well with a blend of sandalwood and cinnamon or else with frankincense for mid-winter comfort.

Winter is also an excellent time to regain some clarity and sense of purpose. The New Year is the time when you can set realistic goals and targets for the future and try to get your life back on track. Making lists and writing things down is a good idea. Light a candle and burn some incense as you inwardly reflect. There are several suitable fragrances to choose from. Dammar resin is transparent and shaped like icicles. Sometimes it is shot through with a yellowish or reddish tint. Dammar means "light" in Malaysia, alluding to the resin's powerful refractive qualities. It is sometimes called "cat's eye resin" in the West as its delicate lemony fragrance brings light where there is darkness and confusion. Dammar is said to aid clairvoyance, which literally translates from the French as "clear seeing". It combines well with other lemony fragrances such as elemi and lemongrass.

Another resin that forms into stalactites is sandarac. It ranges in colour from lemon yellow to burnt amber and has a shiny, glass-like surface when broken in half. Its aroma is light and balsamic, similar to frankincense or high-quality copal resin. Sandarac is clarifying and will help strengthen resolve when you find it difficult to stick to your plans. Burned by itself it is very smoky, so is best used with other fragrances, such as coriander and benzoin. Other clarifying aromas are fresh herbaceous fragrances, such as rosemary, sage, juniper and thyme. These may be used alone or combined with mastic or any other suitable resin for a fuller blend.

Sharp contrasts and warm, rich colours evoke winter. Scents are

spicy and invigorating, clear and illuminating or decadently self-indulgent.

recipes

Aromatics are rather like food – there are endless ways of combining them to suit all sorts of moods and occasions. The following recipes are to get you started, but do not be afraid to branch out and substitute your own ingredients.

In all the following recipes, the quantities may be increased if you wish, but keep the proportions in the same ratio. If you are unable to obtain an ingredient, then try substituting something else with similar fragrance properties.

INCENSE BLENDS

These suggested incense recipes use a selection of ingredients that you should be able to find fairly easily.

REFRESH

Morning refresher

Wake up your senses with fresh citrus scents.

YOU WILL NEED

10ml (2 tsp) labdanum (resin or oil)
25ml (5 tsp) mastic
5ml (1 tsp) dried rosemary
5ml (1 tsp) dried lavender
2.5ml (½ tsp) dried lemon thyme
dry peel of a lime
1 drop bergamot essential oil
pestle and mortar

1 Break the labdanum into small pieces. As labdanum is very sticky, this can be difficult. Try keeping it in the freezer and shaving off a little when you need it with a sharp knife. Alternatively, you may purchase it as an oil.
2 Crush the mastic and mix with the labdanum.
3 Crush the herbs and lime peel together until finely ground.
4 Mix the herbs into the sticky resin. Add the essential oil.
5 Form the mix into pea-size pellets and burn on top of charcoal or mica.

RELAX

Egyptian slumbers

This recipe is based on the famous kyphi mix from ancient Egypt. Kyphi was reputed to bring peace and calm.

YOU WILL NEED

50ml (10 tsp) raisins (unsulphured)
a little red wine
20ml (4 tsp) frankincense
5ml (1 tsp) myrrh
5ml (1 tsp) opoponax (sweet myrrh)
10ml (2 tsp) gum mastic
10ml (2 tsp) benzoin (Sumatra)
7.5ml (1½ tsp) sandalwood
2.5ml (½ tsp) juniper berries
2.5ml (½ tsp) cardamom
2.5ml (½ tsp) cinnamon
1.5ml (¼ tsp) calmus root
2.5ml (½ tsp) galangal or ginger
2.5ml (½ tsp) dried rose petals or buds
5ml (1 tsp) soft honey
pestle and mortar/electric grinder
(use only for incense making)

1 Soak the raisins overnight in red wine.
2 Pulverize the resins (frankincense, myrrh, mastic and benzoin) with the pestle and mortar, or in an electric grinder, and put to one side.
3 One by one, take the sandalwood, juniper, cardamom, cinnamon, calmus and galangal and make into powders. (You may have purchased some of these in powder form.)
4 Mix the powders together with the resin powder. Crumble in the rose buds or petals.
5 Drain the raisins and mix together with the powder in the mortar or electric grinder. Add the honey.
6 Take out the mixture; it should be soft, moist and crumbly. Make into pellets and spread them onto a piece of cloth, away from direct heat or sunlight. Leave the pellets for 10–14 days to dry out, turning every few days.
7 Burn the pellets on top of charcoal or mica.

ENERGIZE

Zest and zing

This spicy fragrance is revitalizing and should help to combat burn-out, particularly from overwork.

YOU WILL NEED

10ml (2 tsp) myrrh
5ml (1 tsp) cloves
2.5ml (½ tsp) cinnamon
3 whole black peppercorns
15ml (3 tsp) sandalwood powder
5ml (1 tsp) powdered ginger or galangal root
pestle and mortar

1 One by one, crush the myrrh, cloves, cinnamon and black peppercorns into powdered form.
2 Mix in the sandalwood and ginger.
3 Use the mixture sparingly and burn on charcoal.

CHEER

Uplift the senses

This exhilarating fragrance is relaxing and enlivening. Try it when morale is low.

YOU WILL NEED

5ml (1 tsp) labdanum resin or oil
10ml (2 tsp) mastic
4.5ml (¾ tsp) coriander seeds
10ml (2 tsp) dried orange peel
5ml (1 tsp) dried lemon peel
pinch of saffron
pestle and mortar

1 Break the labdanum into small pieces. As labdanum is very sticky, this can be difficult. Try keeping it in the freezer and shaving off a little when you need it with a sharp knife. Alternatively, you may purchase it as an oil.
2 Crush the mastic into a powder. Repeat with the coriander seeds and then the fruit peels.
3 Mix all the ingredients together, crumbling in the saffron. If you are using labdanum oil, this should be added last.

SOOTHE

Tension easer

This light floral, herbaceous mix will calm and soothe jangled nerves and restore emotional equilibrium.

YOU WILL NEED

10ml (2 tsp) myrrh
5ml (1 tsp) mastic
10ml (2 tsp) dried camomile heads
5ml (1 tsp) dried lavender
5ml (1 tsp) dried marjoram

1 Crush the resins and bind them together.
2 Mix the dried plant pieces together and crush into a powder.
3 Mix the powder into the resin and burn a little at a time on charcoal or mica.

SEDUCE

Arabian nights

Sensuous temptation is at your fingertips with this mellow luxurious fragrance.

YOU WILL NEED

20ml (4 tsp) sandalwood powder
5ml (1 tsp) benzoin (Siam)
5ml (1 tsp) agarwood
5ml (1 tsp) cinnamon
2.5ml (½ tsp) cloves
2.5ml (½ tsp) dried rose petals or buds
1.5ml (¼ tsp) dried lavender flower-heads
pinch of saffron
1 drop rose absolute oil (optional)

1 Pulverize the sandalwood, benzoin, agarwood, cinnamon and cloves separately and then mix them together.
2 Crumble the dried rose pieces, the lavender and the saffron into the powder.
3 Add the rose oil if needed and the incense is ready for burning.

INSPIRE

Be inspired

Free your imagination and creative powers with this mysterious and evocative blend of resins and spices.

YOU WILL NEED

10ml (2 tsp) sandalwood
5ml (1 tsp) benzoin (Siam)
5ml (1 tsp) star anise
5ml (1 tsp) cinnamon or cassia
1.5ml (¼ tsp) cloves

1 Grind all the ingredients individually and then mix together, adding the spices to the resins.
2 Burn a little at a time on charcoal or mica.

DREAM

Vivid dreaming

To stimulate dreaming, burn a little of this incense last thing at night before going to bed.

YOU WILL NEED

15ml (3 tsp) white or gold copal
5ml (1 tsp) night copal
5ml (1 tsp) dried dream herb
2.5ml (½ tsp) dried hops
pestle and mortar

1 Break the copal into small pieces and crush into powder.
2 Crumble the dried hops and dream herb into the mix.
3 Use the incense a pinch at a time on charcoal or mica.

REFRESH

Clear open spaces

This herbal aroma is based on traditional Native American cleansing plants.

YOU WILL NEED

15ml (3 tsp) white sage
10ml (2 tsp) sweet grass
10ml (2 tsp) cedar leaves
5ml (1 tsp) desert mugwort

1 Cut up the herbs and mix them together.
2 Burn a little at a time on charcoal or mica.

HEAL

Balance and harmony

In times of distress, this healing blend will help restore harmony.

YOU WILL NEED

7.5ml (1½ tsp) frankincense
7.5ml (1½ tsp) myrrh
7.5ml (1½ tsp) mastic
2.5ml (½ tsp) storax
2.5ml (½ tsp) calmus
pinch of cinnamon (optional)

1 Pulverize the resins and mix together.
2 Cut up the calmus and add to the resins.
3 Mix in the cinnamon, if using, and burn over charcoal or mica.

TRANSFORM

Angel wings

Float away on this delicate fragrance and journey to the stars.

YOU WILL NEED

10ml (2 tsp) frankincense
10ml (2 tsp) mastic
5ml (1 tsp) dammar
5ml (1 tsp) sandarac
pestle and mortar

1 Crush the resins into small pieces and mix together.
2 Burn sparingly over charcoal or mica.

RITUAL

Sacred silence

Hear the unfolding of your soul as you "listen" to the fragrance.

YOU WILL NEED

10ml (2 tsp) frankincense
5ml (1 tsp) cinnamon
5ml (1 tsp) sandalwood powder
5ml (1 tsp) agarwood

1 Pulverize the frankincense and crush the cinnamon. Mix together.
2 Add the sandalwood powder.
3 Shave slivers of agarwood into the mix.
4 Burn on Japanese charcoal or mica.

SPRING

Spring cleanse

This uplifting and detoxifying blend should clear away any winter cobwebs.

YOU WILL NEED

10ml (2 tsp) mastic
20ml (4 tsp) juniper berries
10ml (2 tsp) white sage
2 drops eucalyptus essential oil

1 Pulverize the mastic resin.
2 Crush the juniper berries and white sage to a fine powder.
3 Mix the powder with the resin and add the eucalyptus oil.
4 Use sparingly over charcoal or mica.

SUMMER

Summer solstice

Burn this mix outside to celebrate summer solstice and connect with nature.

YOU WILL NEED

15ml (3 tsp) frankincense
5ml (1 tsp) myrrh
5ml (1 tsp) elecampane
2.5ml (½ tsp) mugwort

1 Pulverize the resins and mix together.
2 Cut the elecampane and mugwort into small pieces and add to the mix.
3 Burn a pinch at a time on charcoal or in an incense burning vessel.

AUTUMN

Mellow mood

Capture the richness of autumn with this warm and evocative fragrance. This incense blend uses rich, spicy ingredients such as cinnamon and sandalwood to create a celebration of autumn plenty and earthy fragrances.

YOU WILL NEED

10ml (2 tsp) myrrh
5ml (1 tsp) benzoin (Siam)
10ml (2 tsp) sandalwood powder
2.5ml (½ tsp) cinnamon
pinch of saffron

1 Pulverize the resins individually and then mix them together.
2 Add the sandalwood, crushed cinnamon and saffron.
3 Use the incense sparingly and burn on some charcoal.

WINTER

Winter nights

This soft, balsamic aroma is both comforting and sensuous. It is perfect for burning on a cold winter's night in order to create a warm and inviting atmosphere, particularly if you are welcoming guests into your home.

YOU WILL NEED

15ml (3 tsp) gold copal
5ml (1 tsp) night copal
5ml (1 tsp) myrrh
5ml (1 tsp) balsam of tolu
(this is optional)
2.5ml (½ tsp) crushed vanilla pod
pinch of cinnamon powder
pestle and mortar

1 Grind the resins one by one and then mix them together.
2 Add the vanilla and cinnamon powder to the mix. If you are not using the tolu balsam, then you can increase the amount of vanilla to 5ml (1 tsp) instead.
3 Burn in small amounts, a pinch at a time, on charcoal or mica.

OTHER METHODS

There are many other ways of using fragrance around the home. The following recipes are to give you some ideas.

LAVENDER SACHETS

Sachets are simple to make and are a good way to scent linen and clothes. The ingredients can be as varied as you wish.

YOU WILL NEED

2 pieces soft cotton fabric, approximately 10cm (4in) square
needle and thread
25ml (5 tsp) dried lavender heads
5ml (1 tsp) dried rosemary
5ml (1 tsp) dried thyme

1 Make a 5mm (¼ in) hem along one side of each of the fabrics.
2 Match up the two hems and lay the two pieces of fabric on top of each other.
3 Sew a seam around the remaining three sides and turn inside out.
4 Fill the sachet with the dry ingredients and stitch the open side neatly together.

POT-POURRI MIXES

You can make your own pot pourri from a range of natural ingredients, including herbs, spices, flowers and essential oils.

Scented pine cones

Collect pine cones in the autumn and impregnate them with aromatic oils. Use them in cupboards and around the home.

YOU WILL NEED

Pine cones
Old washing-up bowl
Essential oils: choose from cedarwood, ginger, cloves, cinnamon or patchouli.

1 Leave the pine cones to dry out in a warm place for a couple of weeks.
2 Fill the bowl with water and add the essential oils. Use up to 25 drops of oil to every 150ml (¼ pt) of water.
3 Soak the cones overnight in the water.
4 Drain the water and dry off the cones.
5 Refresh the cones by resoaking.

Spicy citrus pot pourri

This colourful pot pourri will fill the air with its spicy fragrance. It can also be used as a sachet filler.

YOU WILL NEED

200ml (1 cup) dried orange peel
120ml (½ cup) dried lemon peel
200ml (1 cup) dried bayleaves
200ml (1 cup) dried lemon thyme
200ml (1 cup) dried lavender flower heads
120ml (½ cup) dried lime blossoms
2.5ml (½ tsp) grated nutmeg
2.5ml (½ tsp) cloves
1.5ml (¼ tsp) cinnamon
30ml (2 tbsp) ground orris root
2 drops lavender oil
2 drops bay oil
2 drops geranium oil
1 drop orange oil
Airtight container

1 Take the orange and lemon peel and break into pieces. Crush the rest of the dry ingredients into smaller pieces and mix together. Add the peels.
2 In a separate bowl, mix the spices and orris root. Add the essential oils and combine with your fingers, as though rubbing fat into flour.
3 Add the spice mix to the dry ingredients and mix together.
4 Place the pot pourri in the container and leave for 4–6 weeks, shaking it from time to time.
5 Decant the mix into decorative bowls and display as required.

Rose and vanilla pot pourri

This pot pourri looks good when layered into a clear glass container. An airtight storage jar is ideal. The delicate scents mingle to give a sumptuous fragrance.

YOU WILL NEED

½ vanilla pod
10ml (2 tsp) ground cinnamon
2.5ml (½ tsp) ground cloves
30ml (2 tbsp) orris root
5 drops rose absolute oil
5 drops rosemary oil
2 drops geranium oil
1 drop patchouli oil
750ml (3 cups) dried rose petals
750ml (3 cups) dried mint leaves
120ml (½ cup) dried sage
750ml (3 cups) dried rosebuds
120ml (½ cup) dried rosemary
Clear glass storage jar or other airtight glass container

1 Chop the vanilla and mix with the spices and orris root.
2 Mix the essential oils together in a glass shaker bottle and shake well to mix.
3 Put a layer of rose petals in the glass container. Sprinkle a little of the orris mix and a drop or two of the essential oils over the petals.
4 Mix the mint and sage together and put a layer over the rose petals. Sprinkle with the orris and essential oil mixes as before.
5 Add a layer of rosebuds, followed by the rosemary, sprinkling the orris and oils over each layer.
6 Repeat this layering process, finishing with a final layer of rose petals.
7 Put the lid on the jar and leave in a dark place for 6 weeks before use. This allows the fragrances to develop.

Summer spice pot pourri

This pot pourri looks good when mixed into a bowl of coloured pebbles.

YOU WILL NEED

200ml (1 cup) dried lavender
120ml (½ cup) dried rose petals
120ml (½ cup) dried lemon balm
5ml (1 tsp) allspice
5ml (1 tsp) mace
10ml (2 tsp) ground orris root
2 drops rose oil

1 Mix the flowers and herbs together.
2 Add the spices, orris root and rose oil and shake well.
3 Put the mixture in an airtight container and leave for 2–3 weeks, shaking occasionally. Use as required.

ROOM VAPORIZERS

The following combinations are for vaporizing in an oil burner. Vaporizing essential oils is a quick and effective way of adding fragrance to a room. The mixtures can also be added to a mist-spray bottle.

Unwind

3 drops frankincense
2 drops orange
1 drop myrrh

Chill out

3 drops geranium
2 drops lavender
1 drop camomile

Freshen up

3 drops lemongrass
2 drops orange
1 drop basil

Purity

3 drops eucalyptus
2 drops juniper
1 drop rosemary

Inner calm

4 drops lavender
2 drops neroli
1 drop petitgrain

Breathe easy

3 drops peppermint
1 drop eucalyptus
1 drop lemon

Exotic

3 drops neroli
2 drops ylang-ylang
1 drop jasmine

Spirit

3 drops sandalwood
2 drops frankincense
1 drop rose absolute

Vision

4 drops lemongrass
1 drop nutmeg
1 drop cardamom

suppliers

United Kingdom

Arcania
17 Union Passage, Bath BA1 1RE
Tel: 01225 335233
Native American smudge bundles, loose herbs, incense powders and resins, handmade incense sticks and essential oils.

Aromatic Moods
Email: info@aromaticmoods.co.uk
www.aromaticmoods.co.uk
Essential oils, incense sticks and cones, pot pourri, candles and room scents. Also giant incense sticks, ideal for outdoor use.

Baieido
www.baieido.com
Japanese incense using all natural ingredients in a wide variety of styles

Holistic Shop
PO Box 46, Attleborough, Norfolk
NR17 2WB
Tel: 01953 456897
Email: admin@holisticshop.co.uk
www.holisticshop.co.uk
Indian, Nepalese, Tibetan, Native American and Japanese incense sticks and cones.

Neals Yard Remedies
15 Neals Yard, Covent Garden,
London WC2H 9DP
Tel: 020 7379 7222
www.nealsyardremedies.com
Loose herbs, essential oils and incense resins.

Pan's Pantry
www.panspantry.co.uk
Loose resins, oils, woods, herbs and spices.

Silk Road's End
www.silkroadsend.com
email: sales@silkroadsend.com
Specialist supplier of incense resins and equipment. Also premium-quality agarwood and Japanese charcoals and handmade Japanese incense sticks.

United States

Dreaming Earth Botanicals
P.O. Box 386
Hiawassee, GA 30546
Tel: (800) 897-8330
Email: info@dreamingearth.com
www.dreamingearth.com
Essential oils and aromatherapy products.

Incense Galore
1205 North Idaho Street
Post Falls,
Idaho 83854
Tel: (866) 259-3253
www.incensegalore.com
Incense oils, sticks and cones.

J.Crow Company
P.O. Box 172
New Ipswich,
NH 03071
Tel: (800) 878-1965
Email: jcrow@jcrow.mv.com
www.jcrows.com
Essential oils, fragrances, herbs, spices, roots, barks, incense sticks and cones.

left *The smell of coffee is stimulating. Drinking coffee can help keep you awake when you are working late into the night.*

right *The sweet, enticing smell of vanilla can be used to create a calming and relaxed atmosphere.*

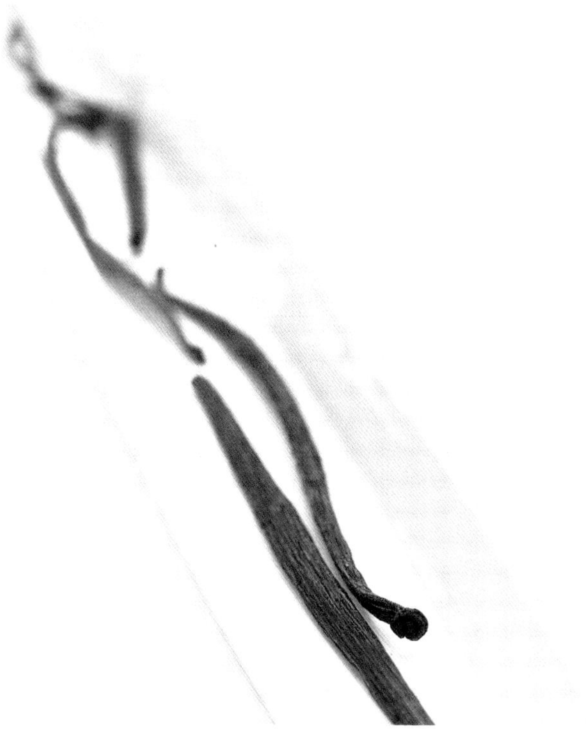

Nutraceutic, Inc.
P.O. Box 358331, Gainesville, FL 32635
Tel: (352) 371-3735
Email: info@nutraceutic.com
www.nutraceutic.com
Incense from around the world.

Scents of Earth
PO Box 859, Sun City CA 92586.
Tel: (800) 323 8159
www.scents-of-earth.com
Aromatic resins, woods and herbs; incense starter kits; ceremonial incense supplies.

Shoyeido Incense
www.shoyeido.com
Tel: (800) 786 5476
Producer of fine-quality Japanese incense.

Canada
Birdie's Nest
90 Park Avenue East
Chatham, ON N7M 3V4
Tel: (519) 354-4040
www.birdiesnest.on.ca
Incense and essential oils.

Green Valley Aromatherapy Ltd.
4988 North Island Highway
Courtenay, BC V9N 9H9
Tel: (250) 334-4836
Email: greenvalley@57aromas.com
www.57aromas.com
Essential, perfume and carrier oils and bases.

Australia
221 Kerr Street,
Fitzroy, Melbourne 3065
Victoria
Tel: (61) 3 9486 9688
Fax: (61) 3 9486 9388
www.inessence.com.au
Essential oils, vaporizing and accessories.

Sunspirit
P.O. Box 85
Byron Bay
New South Wales 2481
Tel: (61) 2 6685 6333
Fax: (61) 2 6685 6313
Email: sunspirit@sunspirit.com.au
www.sunspirit.com.au
Essential oils, carrier oils, oil burners and vaporizers.

New Zealand
Rivendell
Westfield Shopping Towns, Auckland
Glenfield Shopping Centre, tel: (09) 442-1640; Manukau Shopping Centre, tel: (09) 263-4540; St Lukes Shopping Centre, tel: (09) 846-8277; West City Shopping Centre, tel: (09) 836-9019
Magical and mystical, spiritual incense, oils and crystals.

The House of Pamela
33 Rangeview Road,
Sunnyvale
Henderson
Auckland
Tel: (09) 835-1165
email houseofpamela@slingshot.co.nz
100% pure essential oils, reasonably priced.

index

Acknowledgements

Authors' acknowledgements

With thanks to Calmia; Candles; CBC; Grange; Kenneth Turner Candles; Khalid Miller; Molton Brown; Pure Pr; and Sophie at Light Locations.